Basic
Mathematics
Literacy Book

And Study Companion

P. K. Prah

Basic Mathematics Literacy Book
And Study Companion
All Rights Reserved.
Copyright © 2016 P. K. Prah
v2.0

Cover Photo © 2016 P. K. Prah. All rights reserved - used with permission.

Outskirts Press, Inc.
http://www.outskirtspress.com

ISBN: 978-1-4787-7161-6

Library of Congress Control Number: 2016901855

Outskirts Press and the "OP" logo are trademarks belonging to Outskirts Press, Inc.

PRINTED IN THE UNITED STATES OF AMERICA

TABLE OF CONTENTS

Chapter 1 – Getting Started

1.1 Introduction

Our objective for creating this material is to make basic mathematical concepts easy to understand. The material is intended for:

- People who are interested in improving their literacy in mathematics.
- People who have struggled, or are struggling, with basic mathematics.
- People with a specific objective, such as to acquire a General Educational Development (GED) diploma.

We don't always do mathematics in a formal setting. When we work out how much to leave for tip after dinner in a restaurant, we are doing mathematics. When we figure out how much weight we need to lose to hit a specific target, we are doing mathematics. The clerk who counts inventory at a store and determines when it is time to replenish inventory levels does mathematics. We all do mathematics as a natural part of our lives. As such, we all benefit from a level of literacy in mathematics.

Mathematics is a discipline that uses logic and symbols to create a structure that allows us to analyze and solve problems in a consistent way. In that sense, mathematics has a language of its own. Mastering mathematics requires proficiency in that language.

The following are examples of some of the symbols and words we will encounter in this short study we are undertaking. You are likely familiar with them from an earlier introduction to arithmetic, which is a branch of mathematics.

Some arithmetic symbols:

+ (addition),

− (subtraction),

× (multiplication),

÷ (division).

We call these symbols **operators**. We use them to perform arithmetic **operations**.

An arithmetic operation involves two numbers, called **operands**, and one operator. For example: **2 + 3** is an arithmetic operation, involving the addition operator and the numbers, or operands, **2** and **3**.

Already, we are employing words like **operator**, **operation**, and **operands**. They are examples of what we mean by the language of mathematics. Pay attention to the language and symbols of mathematics as we encounter and discuss them. Make an effort to understand their use. If you do that, you will find the material easy and interesting.

We will focus on basic concepts, and point out shortcuts as they apply to a topic. However, we will not dwell on shortcuts for solving problems. If you get grounded in the basics, you will recognize the opportunities for using shortcuts during problem solving.

You know the expression: practice makes perfect? Well, it especially applies to mathematics, even at basic a level as ours. After we cover a topic, complete the related review problems.

If you are going after a GED diploma, get a GED practice book from your local library, then work on as many more practice problems as you can from the GED practice book. We have provided a self-assessment test in the back of the book for additional practice.

Answers to both the review problems and the self-assessment test can be found in the back of the book.

Before we get into things, let us re-familiarize ourselves with numbers.

1.2 A Brief Look at Numbers

We are all familiar with numbers at some level. After all, we learned to count in early childhood using our fingers or things around the house. Here we will examine numbers more closely using the concept of the Number Line.

Figure 1: The Number Line

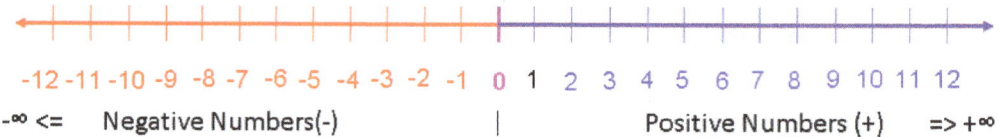

Figure 1 is a pictorial of the number line. A line is made up of points. Each point on the number line represents a **real** number.

On the number line, we have indicated a few points that are occupied by **whole numbers**. Another name for a whole number is **Integer**.

Zero is the center of the number universe. Numbers greater than zero are **positive**. Numbers less than zero are **negative**. On the number line we show positive numbers to the right of zero, and negative numbers to the left of zero. Zero itself can be grouped with positives or negatives depending on the situation.

The sign for a positive number is (+), and the sign for a negative number is (−). Usually, we do not write the sign of a positive number, but we must write the sign of a negative number to differentiate it from a positive number.

Think of a number as consisting of two parts: a "**bigness**" and a **sign**. **Bigness** indicates the number's distance from zero on the number line. **Sign** indicates whether the number is positive or negative. For example, the numbers (**-2**) and **2** have the same bigness, but the first is negative and the second is positive. In mathematics the name for the bigness of a number is **Absolute Value**.

A negative number is not a mystery! It is simply a number that is less than zero. You most likely have heard the weatherman say that the temperature is 50 degrees in one place and minus ten (-10) degrees in another. When the temperature is above zero, it is positive, like 50 degrees. When the temperature is below zero, it is negative, like (-10) degrees.

Similarly, a company can have a deposit of $2 million in a bank, or it can owe the bank $2 million because it took a loan. On the bank's books the account balance of the company will be recorded as **$2** million in the first case and $**-2** million in the second case.

Note that when we start at any number on the number line and move to the right, we **increase** our position by the number of units we move. That is, we **add** the units we move to our starting number. Similarly, when we start at any number and move to the left, we **decrease** our position by the number of units we move. That is, we **subtract** the units we moved from to our starting number.

Numbers that represent actual numeric values, like the numbers we show on the number line, are called **numerals**. We don't always work with numerals. Often we work with numbers whose numeric values are unknown. We call such numbers **variables**, and assign them a letter name. Any letter of the alphabet will do, but '**x**' is one of the letters most commonly used to designate a variable.

The concept of a variable should not surprise us. In real life we occasionally encounter a reference to person **X** in the newspaper. We don't know whether person **X** is male, female, young, or old. But that does not make **X** any less a person. With enough information we can tell who person **X** is exactly.

By the same token, a variable is as much a number as the number **6**. We use variables in mathematical operations in much the same way we use numerals. Given enough facts we figure out the value of a variable exactly.

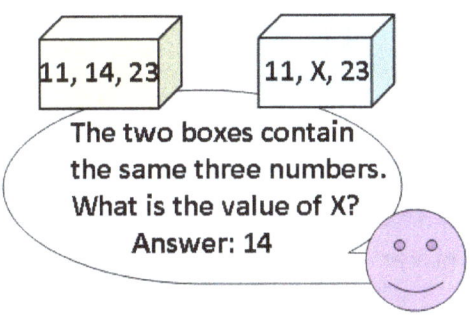

In the graphic above, **x** is a variable. Based on the information provided, we easily determine that **x** equals **14**. If the middle number in first box had been **20**, instead of **14**, the value of **x** would be **20** instead of **14**.

The facts of the problem we face help us determine the numeric value of a variable. As the facts change, so does the value that the variable takes on; that's why it is called a variable.

We write real numbers in many forms: as whole numbers, as shown in *Figure 1*, as fractions, as decimals, as exponentials, and so on. We will discuss these other forms later.

Whole numbers, or integers, are multiples of the number **1**. We illustrate this in the graphic below. We can count whole numbers using our fingers.

Integers are multiples of 1

| 1 | 2 | 3 | 4 | 5 | 6 |

Since integers are multiples of **1**, each integer is one unit apart from the integer next to it. When we start at zero on the number line and move one unit to the right, we get to the integer (**1**). When we move one more unit to the right, we get to the integer (**2**), and so on. The numbers get **bigger** as we continue rightward. We can keep going to the right on the number line without ever running out of numbers. This means that there is an infinite count of positive numbers.

Similarly, when we start at zero and move to the left one unit, we get to the integer (**-1**), called **negative** (**1**). One more unit to the left gets us to the integer (**-2**), and so on. The numbers get **smaller** as we continue leftward; so the number (**-2**) is **1** less than the number (**-1**). Similarly, the number (**-3**) is **1** less than the number (**-2**). We can keep going to the left without ever running out of numbers. This means that there is an infinite count of negative numbers as well.

Between consecutive integers, positive or negative, there is another infinite list of numbers. For example, between 0 and 1, we have every number that is greater than 0 and less than 1. All the following numbers lie on points between 0 and 1: $\{^1/_{1000},\ ^1/_{999},\ ^1/_2,\ ^3/_4,\ ^7/_8,\ ^{11}/_{12},\ ^{999}/_{1000},\ ...\}$. These numbers are examples of **fractions**.

Similarly, between 1 and 2, we have every number that is greater than 1 and less than 2. All the following numbers lie on points between 1 and 2:

$\{1^1/_{1000},\ 1^1/_{999},\ 1^1/_2,\ 1^3/_4,\ 1^7/_8,\ 1^{11}/_{12},\ 1^{999}/_{1000},\ ...\}$. Numbers consisting of integers and fractions are called mixed numbers. We will cover mixed numbers when we discuss fractions.

The same situation is true for consecutive negative integers.

The symbol (∞) stands for **infinity**, the largest possible number. The smallest possible number is (-∞). Infinity, however, is an elusive number because it is so big we can never reach it. When someone says they have an infinite number, we can add **1** and beat theirs by **1**. This can go on indefinitely.

1.3 Even, Odd, Prime Numbers

Numbers that can be divided by **2** without a remainder are called **even** numbers. The term **divisible** means <u>divides without a remainder</u>. So, **even** numbers are divisible by **2**.

Look at the following string of randomly selected numbers:

2, 4, 6, 12, 36, 124, 8968.

You can see that they are all divisible by **2**. They are all **even**. In fact, any whole number which ends in an even digit: **0, 2, 4, 6, 8**, is even.

A number that is not divisible by **2** is an **odd** number. Look at the following other string of randomly selected numbers:

1, 3, 5, 7, 11, 23, 27, 81, 103, 899, 2311.

When we divide any of them by **2**, we get a remainder of **1**.

A prime number is a number that is only divisible by **itself** and the number **1**. Examples of prime numbers are:

1, 2, 3, 5, 7, 11, 13, 17, 19, 23, 31, 37.

Prime numbers are odd, with the exception of the number **2**. However, not all odd numbers are prime.

Now that we are more familiar with numbers, let us move on to using numbers in arithmetic operations. Initially we will examine basics involving individual arithmetic operators. Also, we will consider only integers. From that foundation, we will progress to more complex operations involving fractions, decimals, variables, and so on.

A note before we proceed. We use => in many of the graphics as shorthand for the word **becomes**. It is not a mathematical symbol.

Chapter 2 – Arithmetic Operations

2.1 Our Focus

In this chapter, we will focus on the basics of arithmetic operations, taking one operator at a time. We will discuss the properties of each operator and the related mathematical terminology. We will also point out English language phrases that are commonly used in connection with each operator.

2.2 Addition +

Consider the following arithmetic expression:

> **Example 1**: 2 + 3.

It involves the **addition** operator and the numbers **2** and **3**. The numbers being added are called **operands**. Adding the operands **2** and **3** is like placing two apples in one pile and three apples in another pile and then putting the two piles together. It results in one pile of **5** apples.

In arithmetic, we indicate the result as follows:

> 2 + 3 = 5.

The symbol: **=** (equal to)

indicates that when you add **2** and **3**, the result *equals* **5**. Another way to say this is: the **sum** of **2** and **3** **equals 5**. "Sum" is another English language word that means **addition**. <u>Sum of</u> **2** and **3** means <u>addition of</u> **2** and **3**.

Note that **2 + 3** and **3 + 2** both equal **5**.

When we add any two numbers, it does not matter which one comes first. This property of addition is referred to as the **Commutative Property**. It says that either of the two numbers, or operands, being added can be placed to the left or the right of the addition sign without affecting the result of the addition.

The following is the general way the Commutative Property of addition is stated: If a + b = c, then b + a = c. The letters **a**, **b**, and **c** represent any three numbers we can think of. In **Example 1**, **a** takes on the value **2**, **b** takes on **3**, and **c** takes on **5**. The Commutative Property also applies to Multiplication, but **not** to Subtraction and Division, as we will see later.

We string together numbers and operators to create an **expression**. *Example 1* is an expression, even though it is a very simple one. We **evaluate** an expression in **steps** involving **one operator** and **two** operands at a time. The word **evaluate** means **find the value of**.

When we string together two or more addition operators, another property comes into play. It is called the **Associative Property of Addition**. (This property also applies to Multiplication, but not to Subtraction and Division, as we will see later.) The associative property of addition says that when we string together two or more addition operations, we can perform the operations in any order.

Here is the mathematical statement of the associative property. We use the parentheses simply to indicate which one of the two operations we do first, or which two operands we add first. The associative property says that the order in which we do the addition does not make any difference.

$$a + (b + c) = (a + b) + c.$$

Consider the following expression:

Example 2: 3 + 4 + 5 + 1

We evaluate the expression by performing the operations in steps, two operands at a time. Each step produces an intermediate result which we use in the next step. (We will go from **left** to **right**.) These are the steps:

- we add **3** and **4** and get **7** (result of that step),
- then we add **7** and **5** to get **12** (the result of that step),
- then we add **12** and **1** to get **13** (final result).

Here is another look at these steps:

3 + 4 = 7

7 + 5 = 12

12 + 1 = 13

The graphic below illustrates the steps.

You can see that we worked with two operands at a time, and in some steps one operand was the result of the previous step.

Even though we evaluated the expression in *Example 2* going from left to right, *we did not have to*. The Associative Property says that in an expression involving only addition, we can add the operands in any order and arrive at the same result. Let us evaluate the same expression, this time going from right to left.

>   *Original expression*: $3 + 4 + 5 + 1$

Going from right to left, we add **1** and **5** to get **6**, then we add **6** and **4** to get **10**. Finally, we add **10** and **3** to get **13**. You can see that the order in which we performed the addition did not affect the final result.

2.2.1 Review Problems

Evaluate (find the value of) the expressions and solve the word problem. Do not use a calculator.

i. $23 + 14 + 3 + 8 =$

ii. $1 + 2 + 3 + 4 + 5 + 6 + 7 + 8 + 9 + 10 =$

iii. $101 + 38 + 72 + 96 =$

iv. $1121 + 11 + 5 + 10 =$

v. $10 + 20 + 30 + 40 + 50 + 60 + 70 + 80 + 90 + 100 =$

vi.   Three boys: Jim, Rahul, and Rob went fishing. Rahul caught 18 fish. Jim and Rob each caught 12 fish. How many fish did the three boys catch together?

2.3   Subtraction –

Consider the following arithmetic expression:

>   *Example 1*: $3 - 2$

It is an operation involving the **subtraction** operator and the numbers **3** and **2**. It says: **subtract 2 from 3**. The graphic below illustrates this, and points out some of the different English language phrases we use to indicate subtraction. Subtracting **2** from **3** is like removing **2** apples from a pile of **3** apples.

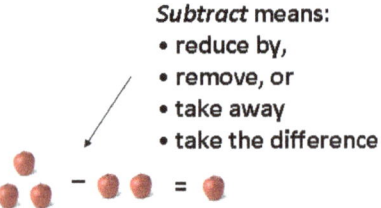

Subtract means:
• reduce by,
• remove, or
• take away
• take the difference

We are left with **1** apple.

In arithmetic, we indicate the result as follows:

3 – 2 = 1.

Again, the symbol: = (equal to)

indicates that when we subtract **2** from **3**, the result equals **1**.

When we subtract one number, say **2**, from another, say **3**, we first write the number we are starting with: **3**, then we write the subtraction symbol: (**–**), followed by the number we are subtracting or taking away: **2**.

Now, consider the following expression:

    Example 2: 2 – 3.

It means subtract **3** from **2**. Here we start with **2** things and take **3** away. Obviously, if you have a pile of two apples and you attempt to take away three apples, you will be **short 1 apple**. Similarly, if you have **2** dollars and you try to buy something that is worth **3** dollars, you will be short **1** dollar.

The result is written as: **2 – 3 = -1**.

The graphic that follows further illustrates this, and once again points out some of the different English language phrases we use to indicate subtraction.

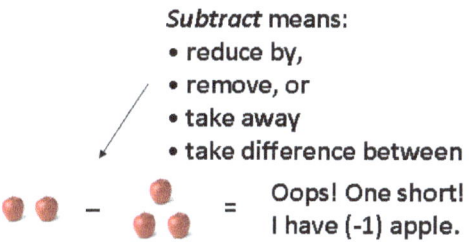

    Subtract **means:**
- reduce by,
- remove, or
- take away
- take difference between

Oops! One short!
I have (-1) apple.

The number (**-1**) is referred to as **negative 1**. As we learned earlier, a negative number is less than zero.

The important point here is that (**2 – 3**) and (**3 – 2**) do **not** produce the same result! Subtraction is **not** commutative, so it is important which operand is placed on the left of the subtraction operator and which one is placed on the right.

Think of it this way. If you have $**3** in your bank account and you write a check for $**2**, there will be $**1** left. This is equivalent to (**3 – 2 = 1**). On the other hand, if you have $**2** in your bank account and you write a check for $**3** to pay for an item, your account will be $**1 short**, or $-1. This is equivalent to (**2 – 3 = -1**).

As we learned earlier, when a number is negative, like (-**1**), we must place the negative sign (-) in front of it to indicate its sign. However, we do not need to write the sign (+) of a positive number, unless we need to do it for emphasis or distinction. When a number does not have the negative sign, we assume it is positive.

The following graphic illustrates positive and negative results.

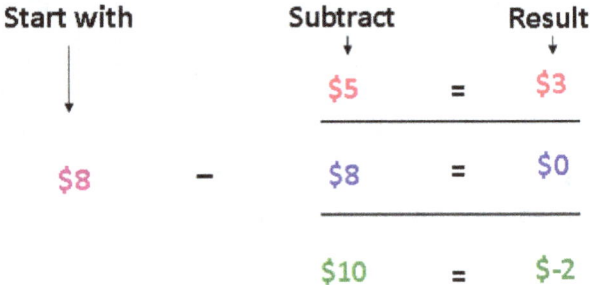

Start with		Subtract		Result
		↓		↓
		$5	=	$3
		_____		
$8	−	$8	=	$0
		_____		
		$10	=	$-2

Think of a positive number as a quantity you **own**, and a negative number as a quantity you **owe**.

Let us take a closer look at the third subtraction in the graphic: **8 – 10** = **-2**.

It is similar to *Example 2*. When we subtract a bigger number from a smaller number, the result is negative. To get the result, we perform the subtraction:

   $10 – 8 = 2$,

then we make the result negative. To emphasize this point, let us consider another case:

   $31 – 17 = 14$,

and: $17 – 31 = -14$.

You can see that result of the second subtraction is the **negative** of the result of the first subtraction.

Did you notice that the symbols we call the **signs** of positive and negative numbers are the same symbols we use for addition and subtraction? This is often a source of confusion for beginners. But it is actually easy to make the distinction between an operation and the indication of a number's sign.

An operation involves two operands. For example, (**3 – 2**) is a subtraction operation; we are subtracting **2** from **3**. Both numbers are positive, that is why we do not indicate their sign. The expression (**3 + -2**) is the addition of two numbers: **3** and **-2**; one of them (**3**) is positive, the other (**-2**) is negative. The symbol (-) in the number (**-2**) is the negative sign, not the subtraction operator.

In the English language, many words and phrases are used to express subtraction.

- Subtract 2 from 3 means (take 2 out of 3): $3 – 2$
- The difference between 3 and 2 means: $3 – 2$
- 3 less 2 means: $3 – 2$
- 3 minus 2 means: $3 – 2$
- 3 reduced by 2 means: $3 – 2$

When we use "difference between", "less", and "minus", "reduced by", we first indicate the number in the pile we are starting with, then we indicate the number we are taking out of the pile. However, when we use "subtract" we name the operands in reverse order.

Now, let us evaluate an expression with a string of subtraction operations. We will use *Example 3* below. The expression tells us to start with a pile of **20** things and successively remove **8**, **6**, and **3**. As long we remove from **20**, we can do the removal in any order we choose. But it will be incorrect to first remove **3** from **6**, or **6** from **8**. This is because the associative property does **not** apply to subtraction. The order in which we perform a string of subtraction operations makes a difference. Going from left to right will always give us the correct result. But as we said above, it is not the only way.

Example 3: 20 − 8 − 6 − 3

We perform the operations in steps. For simplicity, we will go from left to right:

- We subtract 8 from 20 to get 12.
- Then we subtract 6 from 12 to get 6.
- Finally, we subtract 3 from 6.
- The final result is 3.

Here is another look at these steps:

20 − 8 = 12

12 − 6 = 6

6 − 3 = 3

Again, we work with two operands at a time, and in some steps one operand is the result of the previous step. The following graphic illustrates the steps.

Now, let us evaluate the same expression going from **right** to **left** to demonstrate that the results are not the same.

Original Expression: 20 − 8 − 6 − 3

Going from right to left, we subtract **3** from **6** first to get **3**, then subtract **3** from **8** to get **5**. Finally, we subtract **5** from **20** to get **15**. You can see that the two final results are not the same; **15** is incorrect! If we can't remember when the associative property applies, we should simply follow the **left-to-right** rule. It will always get us to the correct result.

Let us evaluate another subtraction expression:

Example 5: 20 – 16 – 7 – 4

Again, we will perform the operations in steps, going from left to right:

We subtract 16 from 20 to get 4.

- Then we subtract 7 from 4. Since 7 is a larger number than 4, we are **short** 3, which is written as (-3).
- Finally, we take 4 from a pile that is already **short** 3; we are **short** 7 altogether.
- The final result is (-7).

Here is another view of the steps we followed:

20 – 16 = 4

4 – 7 = -3

-3 – 4 = -7

Let us take particular note of the final step of **Example 5**:

-3 – 4 = -7

In real life, if you over-draw your checking account by $**3.00**, you have a **negative** account balance of $**3.00**. You owe the bank $3.00, and on the bank's books your account balance is **$-3.00**. If you write another check for $**4.00** without depositing any additional funds to cover your previous over-draft, you owe $**4.00** more, and your account balance is now $**-7.00**.

When we subtract from a quantity, we reduce the quantity by the number being subtracted. If the quantity is already short, then the subtraction creates a bigger short that is equal to the sum of the original short and the amount that is subtracted. In this case we created a bigger short (**-7**) that is the sum of two shorts: (**-3**) and (**-4**). It tells us the following:

-3 – 4 (subtraction of **4** from **-3**)

is equivalent to:

-3 + -4 = - (3 + 4), the same as adding **3** + **4**, and making the sum a short.

The result is (**-7**).

The following graphic illustrates this point.

Our bank account overdrawn by $3		We write a check for $4		Our bank account now overdrawn by $7
Balance $-3	–	$4	=	New balance $-7

2.3.1   Review Problems

Evaluate the expressions and solve the word problems. Do not use a calculator.

  i.   23 – 14 =

  ii.   19 – 1 – 2 – 3 – 4 =

  iii. 101 – 38 – 52 =

  iv.   1211 – 11 – 200 =

  v.   90 – 20 – 30 – 40 =

  vi.   A group of boys have been eating cookies from a jar. First they took 10 cookies. Later they took 13, then 9 more. If the jar had 53 cookies to start, how many cookies are left in it?

  vii. Last year, Mary made $350 a week and saved $90, with the rest going towards her expenses. This year she makes $420 a week and saves $105, with the rest going towards her expenses. What is the difference between Mary's weekly expenses this year and last year?

2.4   Multiplication ×

Think of your earliest introduction to multiplication. You learned that when you multiply two numbers you get a third. Maybe you learned to recite:

2 times 1 = 2,

2 times 2 = 4,

2 times 3 = 6,

2 times 4 = 8, and so on.

You may also have learned that in the English language, the phrase: "**the product of two numbers**" means the same thing as the **multiplication** of two numbers. The product of **2** and **3** is: **2 × 3 = 6**.

Multiplication increases a quantity in jumps. In reality, it is a super-charged addition. Consider the following example:

  Example 1: 4 × 5 = 20

It means that if we put together 5 piles, each pile containing 4 things, the result is one pile of 20 things. The graphic below illustrates this.

$$4 \times 5 \qquad = \qquad 20$$

In the graphic, we add the contents of **5** baskets with each basket holding **4** apples. So, we have a total of **20** apples.

Similarly, if we have **4** baskets with each basket holding **5** apples then we have a total of **20** apples. The next graphic illustrates this second case.

$$5 \times 4 \qquad = \qquad 20$$

We can see from the two cases that: $4 \times 5 = 5 \times 4 = 20$. Recall that earlier we said addition is commutative; that when we add two numbers, it does not matter which number we write first. Multiplication is also commutative. When we multiply two numbers, it does not matter which number we write first.

The following is the general way we express the Commutative Property in multiplication:

If $a \times b = c$, then $b \times a = c$.

As before, the letters *a*, *b*, and *c* represent any numbers we can think of.

Let us look at an expression with a string of multiplication operations:

  Example 2: $4 \times 2 \times 9 \times 1 = 72$

As usual, we perform the operations in steps. We will go from left to right. We multiply 4 and 2 to get 8, then we multiply 8 and 9 to get 72, then we multiply 72 and 1 to get 72. Here is another view of these steps:

  $4 \times 2 = 8$

  $8 \times 9 = 72$

  $72 \times 1 = 72$

The associative property applies to multiplication. It means that any order we choose to multiply the operands will get us to the final result of 72. (You can verify this for yourself by going from right to left.)

2.4.1 Multiplication by One (1) and Zero (0)

In *Example 2* we said **72 × 1 = 72**. When we multiply any number by **1**, the result is the original number. Multiplying by **1** is like making **1** pile of an original quantity. We wind up with the original quantity. The following graphic illustrates **5 × 1**:

5 × 1   =   5

When we multiply any number by zero (0), the result is zero. The size of the original number makes no difference.

2.4.2 Need for Familiarity with Basic Multiplication

Every topic we will cover from this point forward will involve some aspect of multiplication. Therefore, an adequate knowledge of basic multiplication is necessary for continued success.

We have provided a multiplication table in the back of the book. If you had difficulty with the previous discussion of multiplication, you may use the table to refresh your knowledge and for reference when necessary.

2.4.3 Review Problems

Try the following review questions to confirm your understanding of the multiplication operation. Without using a calculator, evaluate the following expressions:

i. $4 \times 14 =$

ii. $19 \times 1 \times 2 \times 3 \times 4 =$

iii. $38 \times 7 \times 2 =$

iv. $11 \times 11 \times 200 =$

v. $20 \times 12 =$

vi.   A man is 35 years old. His daughter is 7 years. What is the product of their ages?

vii.   The area of a rectangle is the product of its length and its width. Area is measured in square units. What is the area in square feet of a rectangular carpet that is 26 feet long and 13 feet wide?

2.5 Division ÷

Division is the inverse or opposite of multiplication. When we divide, we take a pile containing a number of things and break up the pile into smaller piles **of equal size**. The answer we want is the **number of things that each resulting smaller pile contains**.

Let us look at an example:

>**Example 1**: $10 ÷ 5 = 2$

It means we start with a pile containing **10** things and break the pile up into **5** smaller piles of equal size. The result is **2** items in each smaller pile, so our answer is **2**. The following graphic illustrates this.

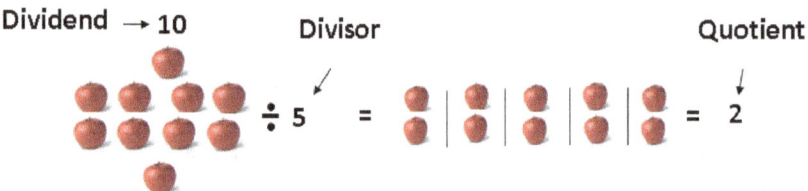

In a division expression, the number to the **left** of the division symbol (÷) is the number being divided. It is called **dividend**. The number to the **right** of the division symbol is called **divisor**; it is the number doing the dividing. The result of the division is called the **quotient**.

In division, just like in subtraction, the order of the operands is important! If we divide a pile of **5** apples into **10** equal smaller piles, there will be **half** an apple in each. So, $10 ÷ 5 = 2$, but $5 ÷ 10 = {}^{1}/_{2}$. Division, like subtraction, is **not** commutative.

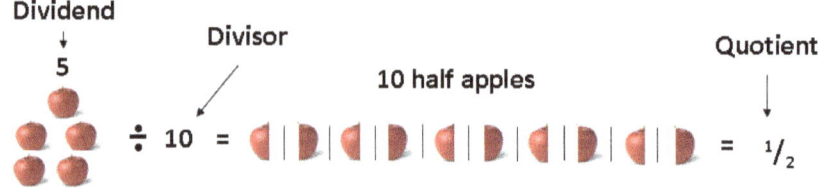

Back in multiplication, we said: $2 × 5 = 10$.

Now, in division, we notice that: $10 ÷ 5 = 2$.

We can see that multiplication reverses the action of division, and division reverses the action of multiplication:

>dividing **10** by **5** gives **2**,

>multiplying **2** by **5** gives **10**.

We capture these relationships in a general statement as follows:

If a × b = c, then b = c ÷ a, and a = c ÷ b.

The graphic below says the same thing in a picture. Like before, *a*, *b*, and *c*, can be any numbers.

If $a \times b = c$

2 × 5 = 10

then $a = c \div b$

2 = 10 ÷ 5

and $b = c \div a$

5 = 10 ÷ 2

In the graphic, we use the values: **a = 2**, **b = 5**, and **c = 10**, to confirm that since **2 × 5 = 10**, we get:

2 = 10 ÷ 5 and **5 = 10 ÷ 2**.

In the English language, we can indicate division in different ways:

10 divided by 2 equals 5:	10 ÷ 2 = 5
2 into 10 equals 5:	10 ÷ 2 = 5
2 divides 10 into 5:	10 ÷ 2 = 5

Here is another common way we write division: $^{10}/_2 = 5$. This is the fraction form. We use the forward slash: **(/)** or a horizontal slash: **(—)** in place of the division symbol. In this form, the **dividend** is the number on **top.** In fractions it is called **numerator**. The number on the **bottom** is the **divisor**. In fractions it is called **denominator.**

Let us evaluate the expression in **Example 3** below, which has a string of division operators. In the expression, **144** is the dividend, while **3**, **6**, and **2** are all divisors. We can divide **144** by **3**, **6**, and **2** in any order we choose. But it will be incorrect to first perform **6 ÷ 2**, or **3 ÷ 6**. The reason is that division is not associative. Similar to what we saw in subtraction, going from left to right ensures the correct answer, but it is not the only way.

Example 3: 144 ÷ 3 ÷ 6 ÷ 2

We go from left to right.

- We divide **144** by **3** and get **48**;
- then we divide **48** by **6** and get **8**;
- now we divide **8** by **2** and get the final result: **4.**

Here is the step-by-step look:

144 ÷ 3 = 48

48 ÷ 6 = 8

8 ÷ 2 = 4

Division is not Associative. We must do successive divisions with the divisors 3, 6, and 2. We can do this in any order, but going from left to right is more straight forward. In this example, if we go from right to left or use any other order that causes a divisor to be divided by another divisor, the result will be incorrect. Let us verify this by going from right to left:

- Divide **6** by **2** to get **3**;
- now, divide **3** by **3** to get **1**;
- finally divide **144** by **1** and the result is **144**!

2.5.1 Division by One (1) and Zero (0)

When we divide a number by (**1**), the result is the original number. Obviously, when we divide a quantity into one pile the resulting pile contains the original quantity. The size of the original quantity makes no difference. For example:

$$5 \div 1 = 5, \text{ and } 100{,}229 \div 1 = 100{,}299.$$

When we divide a number by zero, we get an infinitely large result whose value we cannot determine. This is because in division, as the value of the divisor gets smaller and smaller, the quotient gets larger and larger. For example, when we divide 20 by 20, we get 1; when we divide 20 by 10, we get 2; when we divide 20 by 1, we get 20; when we divide 20 by 0.00001, we get 2,000,000. (We will cover division by 0.00001 in Fractions.) You can see that as we reduce the divisor from 20 to 0.00001, the quotient increases from 1 to 2,000,000. As the size of the divisor approaches zero the quotient becomes so infinitely large that it is **undefined**. As such, we cannot have zero as the divisor in a division.

2.5.2 Review Problems

Work through these problems without using a calculator.

i. $14 \div 14 =$

ii. $27 \div 1 \div 3 \div 3 =$

iii. $65 \div 5 \div 13 =$

iv. $132 \div 11 \div 6 =$

v.   If you divide man's age by 3, you get the age of his son. If the man is 42, what is the age of his son?

vi. A farmer gave 56 apples to 14 children on a school trip to his farm. How many apples did each child get, if they each received the same number of apples?

vii. Tina and Marie start with the dividend 6912. Tina has the number 4 as her divisor and for Marie has 6. Tina goes first; she divides 6912 by 4 and makes her result the new dividend. Then Marie divides the new dividend by 6 and replaces it with her result. They continue this process until one of them reaches a result that is smaller the other's divisor. What is that final result?

Chapter 3 – Negative Numbers

3.1 Our Focus

Most beginners have had little or no introduction to negative numbers. As a result, they approach operations involving negative numbers with some fear. It is our objective to take the mystery out of negative numbers with this discussion.

As we said earlier, negative numbers are as much a part of mathematics as positive numbers. We encountered instances of negative numbers in some of the examples we used in the discussion of **Subtraction**. We are now going to look at negative numbers in more detail.

Again, we will use the number line as our anchor for discussing several examples of addition and subtraction operations involving negative numbers. Since we are already comfortable with positive numbers, we will contrast them with negative numbers when possible.

Figure 1: The Number Line

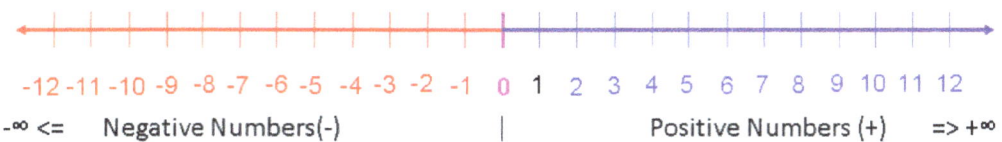

3.2 Adding and Subtracting Negative Numbers

Consider the following example:

Example 1: (-2) + 7 = **5**

We start at the point **(-2)** on the number line. From there we move **7** units to the **right.** We land on the point **5,** which is the result. We move to the right because when we add a **positive** number we **increase** the value we started with. Comparing this result with: **7 – 2** = **5**, we conclude that **(-2) + 7** is equivalent to **7 – 2**, since both equal **5**.

This makes sense. If we have an account balance of $**7** and we write a check for $**2**, we decrease our balance by $**2**. Our balance goes down to $**5**. This is represented by: **7 – 2** = **5**.

Similarly, if we have an account balance of $**-2** (from an overdraft) and we deposit $**7**, the bank uses $**2** from the deposit to cover the negative balance, leaving $**5** in our account. This is represented by: **(-2) + 7 = 5**.

The graphic below illustrates the two cases.

	Starting Account account has $7		We write a check for $2		New Account balance
Case I:	Balance $7	−	$2	=	New balance $5

	Starting Account overdrawn by $2		We deposit $7		New Account balance
Case II:	Balance $-2	+	$7	=	New balance $5

Now, consider the following:

> ***Example 2***: $7 + (-2) = $**5**

This time we are adding the negative number (**-2**). When we add a **negative** number we **decrease** the value we started with. On the number line, we move to the left when we add a negative.

In ***Example 2***, we start at **7** on the number line. We move **2** units to the **left**. We land on the point **5**, which is the result of the addition. Think about this as follows: We have two accounts at a bank, as shown in the graphic below. One account has a balance of $**7**. The other is over-drawn, and has a negative balance of $**-2**. The sum of the two accounts is $**5**.

First account has $7		Second account over-drawn by $2		Net of the of the two accounts is $5
Balance $7	+	$-2	=	New balance $5

This example confirms that **7 + (-2)** is equivalent to **7 − 2**; that is, adding a negative number is equivalent to subtracting the positive of the number. In fact, subtraction is nothing more than the addition of a negative. *This is an important concept. We should remember it.*

We conclude from ***Examples 1*** and ***2*** that **(-2) + 7** and **7 + (-2)** both equal **5**, and are equivalent to **7 − 2**. Earlier we said that Addition is commutative: it does not matter which number comes first. So it should not be surprising that **7 + (-2)** and **(-2) + 7** are equivalent.

Let us look at another example:

> ***Example 3***: $(-2) − 5 = $**-7**.

Again, we start at (**-2**) on the number line. From there we move **5** units to the **left.** We land on the point (**-7**), which is the result. As we learned earlier, we move to the left because subtracting a **positive** number **decreases** the value we started with.

The following is another way to think about this. We start with a **negative** account balance: $**-2**. We write a check for another $**5**. We now have a bigger negative balance totaling $**-7**. The graphic below illustrates this scenario.

Our bank account overdrawn by \$2		We write a check for \$5		Our bank account overdrawn by \$7
Balance \$-2	−	\$5	=	New balance \$-7

Another example:

Example 4: 2 + (-7) = -5

We can easily understand why the result is **-5**. We start on the number line at **2** and move **7** units to the **left**. We land on **-5**. The result is negative because the absolute value ("bigness") of the negative number (**-7**) is larger than that of the positive number (**2**).

Another example:

Example 5: (**-5**) − (**-7**) = **2**.

In this example, the focus is on the part of the expression highlighted in red. What does subtracting a negative mean?

We notice that we get the same result from: **-5** + **7** = **2**.

Thus, we conclude that **-5** − (**-7**) and **-5** + **7** are equivalent. That is, subtracting the negative of a number is equivalent to adding its positive.

Let us examine the claim that **-5** − (**-7**) equals **-5** + **7** further, using an account balance example. First, we will look at **-5** + **7**. Our account balance is \$**-5** because of an over-draft. We deposit \$**7** into the account. We now have \$**2** to our credit. This is equivalent to: **-5** + **7** = **2**.

Next, let us look at **-5** − (**-7**). Our balance of \$**-5** represents five units of debt. We **take away**, or **subtract**, one unit of debt with every dollar we deposit. If subtract seven units of debt when we only owe five units, we will have two units to our credit. This is equivalent to: **-5** − (**-7**) = **2**.

We can also think of the part of the expression: − (**-7**) as a **double** negative. A double negative translates to positive, so − (**-7**) equals + **7**. And we are back to the fact that subtracting a negative number is equivalent to adding the number's positive.

3.3 Multiplying and Dividing Negative Numbers

Again, let us start with an example:

Example 5: **-2** × **5** = **-10**.

We start at zero and take **5** leftward hops of **2** units each. We land on **-10**. Another way to look at this is that a short of **2** duplicated **5** times (**-2** × **5**) results in a short of **10**, or (**-10**). This is illustrated in the graphic below.

$$-2 \times 5 = -10$$

Short Of 2		Short Of 2		Short Of 2		Short Of 2		Short Of 2		Short Of 10
-2	+	-2	+	-2	+	-2	+	-2	=	-10

The expressions **-2 × 5** and **-5 × 2** are equivalent. Two shorts duplicated **5** times **(-2 × 5)** and five shorts duplicated **2** times (-5 × 2) both equal a short of **10**, or (**-10**).

Conclusion: when we multiply a positive and a negative, two numbers with **different** signs, the result is **negative**.

Another example:

> ***Example 6***: -2 × -5 = 10.

When we multiply two negatives, for example: **-2 × -5**, we have a double negative situation. The result is positive: **10.** Remember that when we multiply two positives, for example: **2 × 5**, the result (**10**) is also positive. We conclude that when we multiply two numbers that have the **same** sign the result is **positive**.

Let us consider these results from the perspective of division. From our previous discussion of multiplication and division, we know that if **a × b = c**, then **a = c ÷ b**, and **b = c ÷ a**. As we said earlier, **a**, **b**, and **c** can be any numbers.

Let **a = -2**, and **b = 5**, then: **a × b = -2 × 5 = c = -10**

Therefore: **a = c ÷ b = -10 ÷ 5 = -2** (Case I).

And: **b = c ÷ a = -10 ÷ -2 = 5** (Case II).

We show these results in the graphic that follows.

If → [a] × [b] = [c]
     -2   5   -10

then → [a] = [c] ÷ [b]
        -2   -10   5

and → [b] = [c] ÷ [a]
       5   -10   -2

Conclusion from Case I: When we divide a negative and a positive, the result is a negative. It does not make any difference which of the two numbers is negative. In other words, when we divide two numbers that have different signs the result is always negative.

Conclusion from Case II: When we divide a negative with a another negative, the result is a positive. This is the same result we get when both numbers are positive. In other words, when we divide two numbers that have the same sign, the result is always positive.

3.4 Negative Numbers - Summary

Using the following examples, we summarize what we have learned from operations involving negative numbers as follows:

- **5 – 3 = 2** is equivalent to **5 + -3 = 2**; subtracting **3** means the same as adding (**-3**). Subtraction is really the **addition** of a negative number.

- **-5 + -3 = -8**; using our account balance example, if we add a short (**-3**) to another short (**-5**), we get a bigger short (**-8**). When we add two negatives, the result is negative and its "bigness" is the sum of the "bigness" of the two numbers.

- **5 – (-3) = 8**; subtracting a negative number is equivalent to adding the positive of the number.

- **15 ÷ 3 = 5** and **-5 ÷ -3 = 5**; if both dividend and divisor have the same sign the result is positive.

- **15 ÷ -3 = -5** and **-15 ÷ 3 = -5**; dividing two numbers that have different signs results in a negative.

- **5 × 3 = 15** and **-5 × - 3 = 15**; multiplying two numbers that have the same sign results in a positive.

- **5 × -3 = -15** and **-5 × 3 = -15**; multiplying two numbers that have different signs results in a negative.

3.5 Subtraction <u>is</u> Addition

From our discussion of negative numbers, we now know that subtracting a number is the same as adding the negative of the number. With that in mind, we can always treat subtraction as addition, if we so chose. Consider the subtraction expression **20 – 12 – 4 – 3**. We get the result by performing successive subtractions. We may do this by going from left to right. The result is **1**. If we turn the expression into addition, it becomes **20 + -12 + -4 + -3**. That is, we start with **20** and add a short of **12**, another short of **4**, and another short of **3**. As expected, we are left with **1**. If we have an account balance of $20 and we write three checks for $12, $4, and $3, we are left with $1 in the account. It does not matter in what sequence we write the checks. This brings us back to the observation we made earlier that as long as we subtract **12**, **4**, and **3** each from **20**, we can do it in any order we choose.

Also, one check for $7 has the same effect on our account balance as two checks for $4 and $3, and one check for $19 has the same effect as three checks for $12, $4, and $3. In other words, we can approach the string of subtraction operations as the addition of negatives. We can add all the negatives to get **-19**, then add **-19** to **20**, that is **20 + -19 = 20 – 19 = 1**. (This is the same as adding all the numbers that are being subtracted to get 19, then subtracting that sum from 20.) Sometimes turning subtraction into addition allows us to evaluate an expression in a more efficient way. Also, it helps us to see clearly which numbers are positive and which ones are negative. In this example, it is clear that **20** is positive while the others are negative.

3.6 Review Problems

i. $2 - (-7) + -2 =$

ii. $-3 + 4 + -3 =$

iii. $(12 + -3) \times (2 - -4) + 17 + -4 =$

iv. $(12 - -3) \times (2 + -4) + 17 - -4 =$

v. $(1 - -2) \times 2 - -7 + 11 + -11 =$

vi. Two squirrels, Grey and Red, place an acorn at the (-30) point on the number line. Grey pushes the acorn 3 units in the direction of increasing numbers to the number (-27). Red pushes the acorn 2 units in the opposite direction back to the number (-29). They repeat this process until the acorn **touches** the origin (zero point). How many times does each squirrel push the acorn?

Chapter 4 - Inequalities

4.1 Our Focus

In this chapter, we will discuss the symbols that are used to compare the relative values of numbers. In mathematics we deal with an infinite range of values. We must be able to indicate that one value is bigger than, smaller than, or equal to another.

4.2 Inequality Symbols and their Meanings

The **Inequality** symbols we will discuss are:

> (greater than)

< (less than)

≥ (greater than or equal to)

≤ (less than or equal to)

≠ (not equal to)

The number 4 is less than the number 5. In mathematics we write:

4 < 5.

The symbol (<) means less than.

Similarly, the number 5 is greater than the number 4, so we write:

5 > 4.

You guessed it! The symbol (>) means greater than.

It is easy to differentiate the two symbols from each other. The bigger value "shoots the arrow" towards the smaller value.

If a gym teacher asks every boy in the class who is below 5 feet, 8 inches tall to stand up, we say that the height of every boy standing is **less than** 5 feet, 8 inches. We write: the height of every boy standing < 5 feet, 8 inches. A boy who is exactly 5 feet, 8 inches tall, or taller, is outside the population that the teacher asked for.

By the same token, we can say that the height of every boy sitting down is **greater than or equal to** 5 feet, 8 inches. We write: the height of every boy sitting down ≥ 5 feet, 8 inches.

If the teacher asks every boy who is 5 feet, 8 inches **or less** to stand up, we write: the height of every boy standing up ≤ 5 feet, 8 inches. The symbol (≤) means **less than or equal to**. This time, a boy who is exactly 5 feet, 8 inches is in the population the teacher asked for, and the height of every boy sitting down is > 5 feet, 8 inches.

If we want a set of numbers that are **less than or equal to** 5, the set: {1, 2, 3, 4, 5} qualifies. However, if we want a set of numbers that are less than 5, the set does not qualify. It contains the number 5, which, obviously, is not less than itself.

Note that we say **less than _or_ equal to**, and **greater than _or_ equal to**. We cannot replace _or_ with _and_ because it is not possible to be greater than **and** equal to something at the same time. Similarly, it is not possible to be less than **and** equal to something at the same time.

The symbol (≠) means **not equal to**. It is simple enough; we use it when the values on the left and the right of it are not equal. For example, the number 4 is not equal to the number 5, so we write: 4 ≠ 5.

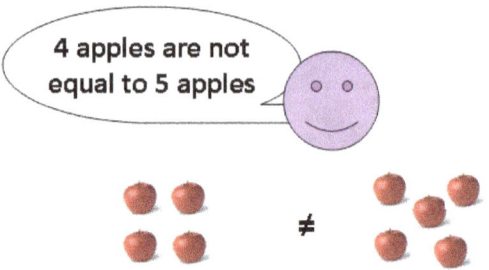

Let us look at some examples involving inequalities. We will use the words **True** and **False** to indicate the validity of each statement.

- 5 < 6   True (**5** is a smaller value than **6**. It is closer to zero on the number line **6**.)

- (8 ÷ 128) < 0   False (The quotient of any two positive numbers is positive, and any positive number is greater than zero.)

- The elements of the set {1,2,3,4} are ≥ 2
  False (One element (**1**) is less than **2**.)

- The elements of the set {1,2,3,4} are ≤ 4
  True (Each element is less than or equal to **4**.)

- -15 > 0   False (**-15** is less than **0**. It is to the left of zero on the number line.)

- -51 < -61   False (**-51** is closer to zero on the number line. It is greater than -**61**.)

- If **a** = **-2**, and **b** = **10**, then **b ÷ a ≠ -5**.
  False (The result of the division equals **-5**.)

4.3 Review Problems

Use **T**-true and **F**-false to indicate the validity of each statement.

i. $134 \div 11 > 13$ ____ T/F

ii. $(2 + 4) \div 2 \leq 2 + 4 \div 2$ ____ T/F

iii. $2(23 - 11) - 20 \geq 2$ ____ T/F

iv. $-2 - -7 = -9$ ____ T/F

v. $-12 + -11 \geq -23$ ____ T/F

vi. $7 + -4 \neq 7 - 4$ ____ T/F

vii.   Mary has a checking account balance of $502. Her friend, Joan, has an account balance of $750. The difference between the two account balances ≥ $247.
                         ____ T/F

Chapter 5 – Absolute Value

5.1 Our Focus

As we learned earlier, we think of a number as consisting of two parts, its "bigness" and its sign. In this chapter, we will discuss the "bigness" of numbers, and introduce the symbol for indicating "bigness".

5.2 Absolute Value

A number's bigness tells us **how far** it is from zero on the number line. The number's sign tells us whether it is positive (on the **right** of zero), or negative (on the **left** of zero) on the number line. The **absolute value** of a number is its bigness.

We have discussed positive and negative numbers extensively because the sign of a number dictates the result of an operation we perform with it. There are times, though, when we are not interested in the sign of a number. We just want to know how big the number is.

The pendulum of a grandfather clock swings to the right a certain distance and then swings to the left the same distance. The clock maker's interest is not in the sign of the distance the pendulum travels; he only cares about making the absolute distance it travels on either side the same.

When we are not interested in a number's sign, we take the number's **absolute value**. The absolute value of a number is indicated with the absolute value symbol: **| |** (two vertical bars).

Let us look at examples:

- The absolute value of **8** is written as **|8|** and it equals **8**.
- The absolute value of (**-8**) is written as **|-8|** and it equals **8**.
-

The absolute value ignores the number's sign. Therefore, a number's negative and positive have the same absolute value.

When we use an absolute value in a calculation, we treat it as a **positive**.

We can take the absolute value of an expression, for example: $|7 - 8 - 4|$. First we evaluate the expression and get the result. Then we take the absolute value of the result. In this case:

$$7 - 8 - 4 = -5,$$

therefore:

$$|7 - 8 - 4| = |-5| = 5.$$

5.3 Review Problems

Solve the following:

i. $|22 - 19 - 8| = $ _____

i. $|-5 \times -5| = $ _____

ii. $|-10| \div -5 = $ _____

iii. $-10 \div |-5| = $ _____

iv. $|-5| \times |-4| = $ _____

v. $|-10| - |-23| = $ _____

vi. $|-2 + -7| = $ _____

Chapter 6 – Order of Operation

6.1 Our Focus

Up to this point, we have focused on the basic arithmetic operators. Also, we have worked with expressions involving only the particular operator under discussion. We are now going to discuss how to evaluate an expression that contains a mix of all the operators we have discussed and two new operators we are going to introduce: **()** and **of**. We will also discuss other ways **()** is used.

6.2 Operators () and of

Before we proceed, let us first get familiar with the operators **()** and **of**.

We know the symbols **()** as **parentheses**. They are also called **round brackets**, or just **brackets**. In our discussion, we will call them brackets. In mathematics they are used as **grouping** symbols. We use them in an expression to indicate an operation, or group of operations, that we want to perform first or perform as a unit. You may recall that we used these symbols in the statement of the Associative Property.

In mathematics **of** is often used as a synonym or replacement for **multiply** or **take the product of**. This use is most common in word problems. For example, **half of 10** means the product of **½** and **10**. When **of** is used in an expression it means the same thing.

6.3 Order of Operation

An expression may have all the arithmetic operators including **()** and **of**. Each operator has an assigned priority. Another word for priority is **precedence**. These relative priorities dictate the order we follow when we evaluate the expression. It is called **Order of Operation**.

Let us start our discussion of order of operation with an example. We have used colors to highlight the operators.

> *Example 1*: $2 \text{ of } 23 - 6 \times (2 + 8) \div 2 + 8 - 7$

How do we go about evaluating the expression in the example? We follow the order of operation rule. Here is how it works:

- Brackets have the highest priority. If there are multiple grouped parts, we do them all first.
- The operator **of** has the next priority. In our example, 2 of 23 is done next.
- Next, we do **division**. We evaluate consecutive division operations from left to right.
- Next, we do **multiplication**.
- Next we do **addition,** then **subtraction**.

Putting it all together, we have: **B**rackets, **O**f, **D**ivision and **M**ultiplication, **A**ddition, and **S**ubtraction. The sequence of first letters spells BODMAS. We use BODMAS as a mnemonic for remembering order of operation.

6.3.1 BODMAS Method

Let us apply BODMAS to *Example 1*. The original expression is:

> **2 of 23 − 6 × (2 + 8) ÷ 2 + 8 − 7.**

Step 1 (**B**): We do **(2 + 8)** to get **10**.

> We replace **(2 + 8)** in the expression with **10**. We now have **2 of 23 − 6 × 10 ÷ 2 + 8 − 7.**

Step 2 (**O**): We do **2 of 23** to get **46**.

> We replace **2 of 23** in the expression with **46**. We now have **46 − 6 × 10 ÷ 2 + 8 − 7.**

Step 3 (**D**): We do **10 ÷ 2** to get **5**. We replace **10 ÷ 2** with **5**. We now have **46 − 6 × 5 + 8 − 7.**

Step 4 (**M**): We do **6 × 5** to get **30**. We replace **6 × 5** with **30**. We now have **46 − 30 + 8 − 7.**

Step 5 (**A**): We add together the numbers being added (the positives). They are **46** and **8**. They add up

> to **54**. Next we add together the numbers being subtracted (the negatives). They are **30** and **7**.

> They add up to **37**.

Step 6 (**S**): We do subtraction: **54 − 37**. The final result is **17**.

A note: The correct use of the BODMAS method hinges on the proper handling of the signs of numbers. In an expression like **-30 − (-2) + 6 + (-4)**, we must be aware that **− (-2)** adds **2**, and **+ (-4)** subtracts **4**. Thus, **2** and **6** are being added and **30** and **4** are being subtracted. We perform **8 − 34** to get **-26**.

Let us work on another example.

> *Example 2*: **24 ÷ 6 × 8 ÷ 4 ÷ 2 × 7 − (8 ÷ 4 − -11 − 10 + 6)**

BODMAS takes us through the following steps:

Step 1: We do the brackets: **(8 ÷ 4 − -11 − 10 + 6)**. We apply BODMAS to the expression inside the brackets. We have division, addition and subtraction. Division has highest precedence. We do **8 ÷ 4** to get **2**. We replace **8 ÷ 4** with **2**. Now we have **2 − -11 − 10 + 6**. The numbers **2**, **11** and **6** are being added. (Remember that **− -11** equals **+11**). We add **2**, **11** and **6** to get **19**, then we do **19 − 10** to get **9**. This completes the brackets. The expression becomes <u>24 ÷ 6</u> × <u>8 ÷ 4 ÷ 2</u> × 7 − **9**. The underlines are for clarity.

Step 2: We do division. We do **24 ÷ 6** to get **4**. We replace **24 ÷ 6** with **4**. Next we do **8 ÷ 4 ÷ 2** to get **1**. The expression becomes **4 × 1 × 7 − 9**.

Step 3: We do multiplication. **4 × 1 × 7** gives **28**. The expression becomes **28 − 9**.

Step 4: We perform **28 − 9** to get **19**.

Let us work on one more example.

> *Example 3*: **4 of 20 − 16 − 1 − 32 ÷ 8 ÷ 4**

Again, BODMAS takes us through the following steps:

Step 1: We do **of** first. **4 of 20** gives **80**. The expression becomes **80 − 16 − 1 − 32 ÷ 8 ÷ 4**.

Step 2: We do division: **32 ÷ 8 ÷ 4** gives **1**. The expression becomes **80 − 16 − 1 − 1**.

Step 3: Only one number is being added. It is the number **80**. The numbers being subtracted add up to **18**.

We perform **80 − 18** to get **62**.

6.3.2 Left-to-Right Method

As an alternative to BODMAS, we can use the **left-** to-**right** method. In the left to right method we modify the order of operation slightly. **()** still has highest priority. We place **of**, **÷**, and **×** on the same level below **()**, and we place **+** and **−** on the same level below **of**, **÷**, and **×**.

So, we still do **()** first, we do **of**, **÷**, and **×** together after **()**, then we do **+** and **−** together.

Here are the steps:

Step1: We do all the brackets and substitute the results.

Step 2: We do **of**, **÷**, and **×**, going from left to right. We substitute the result of each operation as we go.

Final step: We do **+** and **−**, going from left to right. We substitute the result of each operation as we go.

Let us use the left-to-right method to evaluate **Example 2**.

Original expression: **24 ÷ 6 × 8 ÷ 4 ÷ 2 × 7 − (8 ÷ 4 − -11 − 10 + 6)**

Step 1: We do the brackets: **(8 ÷ 4 − -11 − 10 + 6)**. First, we do the division and replace the result. We now have **(2 − -11 − 10 + 6)**. Going from left to right, we add **2** and **11** to get **13**. We subtract **10** from **13** to get **3**. We add **3** and **6** to get **9**. We now have <u>24 ÷ 6 × 8 ÷ 4 ÷ 2</u> × **7 − 9**. Underlines are for clarity.

Step 2: We do division and multiplication, going from left to right. We divide **24** by **6** to get **4**. We multiply **4** by **8** to get **32**. We divide **32** by **4** to get **8**. We divide **8** by **2** to get **4**. We multiply **4** by **7** to get **28**.

Step 3: We do **28 − 9** and get **19**.

If we follow a strict left-to-right discipline, we do not have to worry about which operators are associative and which ones are not. The result will always be correct. However, in some situations we give up the flexibility to perform some operations in a more efficient order.

6.4 Review Problems

Without using a calculator, solve the following problems:

i.   2 of (6 − 2) + 14 ÷ 7 − 5 =

ii.   (39 − 7) ÷ 2 × (9 − 8) =

iii.   15 × (11 + 4) ÷ (17 − 7 + 5) − 13 =

iv. $(20 - 12) \times 2 - 8 \div 4$ of $(21 - 11) + 4 =$

v. $(12 \times 3) - 4 \times (8 - 3 + 4) \div 3$ of $(9 - 7 - 1) =$

vi. $(38 - 17 - 21) \times 3 + 7 \times (5 - 4 - 1) - 19 \times (6 - 5 - 1) =$

vii.   At the end of the day a lobster fisherman, who supplies three restaurants, had a large tub containing 122 lobsters. He shipped 47 lobsters to one restaurant, 35 to the second restaurant, and 20 to the third restaurant. Then he sold half of what was left in the tub to his neighbor. How many lobsters did he keep?

6.5 Other Uses of ()

Earlier we learned that parentheses **()** are used to group operations. We also use them to bring clarity to an expression and other times as shorthand for multiplication.

Compare the following two expressions:

Example 1:

       Expression 1: **5 – -2**

       Expression 2: **5 – (-2)**

Both expressions say the same thing: subtract **-2** from **5**. However, the second expression is clearer because the parentheses make **-2** stand out; the subtraction operator and the negative sign don't run into each other like they do in the first. In this case, we use parenthesis for clarity.

Let us look at the following two expressions for another use of parentheses:

Example 2:

       Expression 1: **2 × (3 + 4)**

       Expression 2: **2(3 + 4)**

Again, the two expressions are equivalent. We are already familiar with the first format. We perform the grouped part by adding **3** and **4** to get **7**, then multiply **7** by **2** to get **14**. The second expression is shorthand for the first one. It looks less complicated because the multiplication operator is not written; it is implied. In a similar way, we can write **2 × 3** as **2(3)** or **(2)(3)**.

6.5.1 The Distributive Property

Let us start with the expression: **2(3 + 4)**. Following the order of operation rule, we would evaluate the expression by first adding **3** and **4** to get **7**, then multiplying **7** by **2** to get **14**, as we did above. But we get the same result if we first multiply each number inside the parentheses with the multiplication factor **2**, then perform the addition, as follows:

       <u>2</u> × 3 + <u>2</u> × 4 = 6 + 8 = 14.

Remember: multiplication before addition.

The graphic below illustrates this.

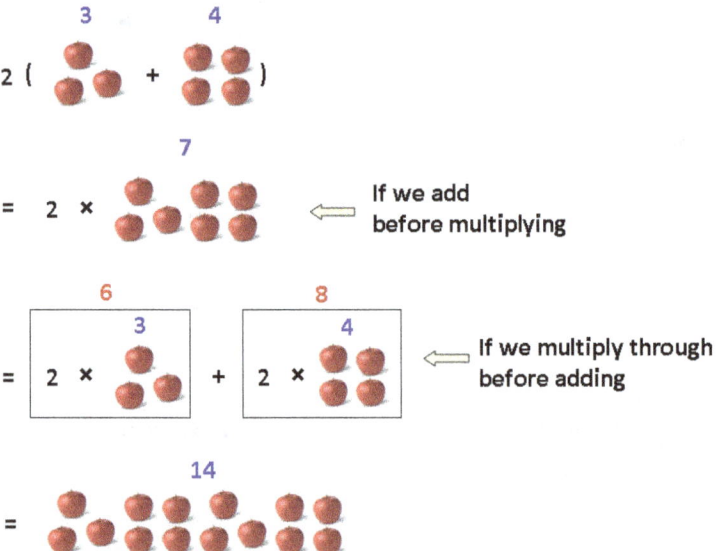

Two approaches, same result!

In mathematics, we state this property in a general way as follows:

a(b + c) = a × b + a × c

and:

(b + c) ÷ a = b ÷ a + c ÷ a

Again, **a**, **b**, and **c** are any numbers. This is called the **Distributive Property of Addition**. In the previous example: **2(3 + 4)**, **a** corresponds to **2**, **b** corresponds to **3**, and **c** corresponds to **4**.

The Distributive Property applies to subtraction as well:

a(b − c) = a × b − a × c

and:

(b − c) ÷ a = b ÷ a − c ÷ a

The Distributive Property says the following: We can follow the order of operation rule by performing the grouped addition or subtraction before multiplying or dividing, or we can first multiply, or divide, each number inside the parenthesis with the multiplication or division factor, then perform the addition or subtraction operation. The result will be the same.

Also, we can do the reverse and go <u>from</u>: a × b + a × c <u>to</u>: a(b + c) and <u>from</u>: b ÷ a − c ÷ a <u>to</u>: (b − c) ÷ a, by factoring out the common multiplication factor or common division factor: a. We will see examples of this in our discussion of Variables.

Here is another example of the application of the Distributive Property involving both addition and subtraction:

> ***Example 1***:   **3**(21 − 10 + 5)

- Doing the grouped part first, we get:

    3(21 − 10 + 5) **=** 3 × 16 **=** 48

- Multiplying through with the factor 3 and then doing the subtraction and addition, we get:

    3(21 − 10 + 5) **=** **3** × **21** − **3** × **10** + **3** × **5** = 63 − 30 + 15 = 48

The Distributive Property is important. We will encounter its application over and over as we move forward.

6.5.2 Review Problems

Evaluate the following expressions.

i. $2(6 - 2) + 4(7 − 5) =$

ii. $2(6) − 2(2) + 4(7) + 4(6 − 5) =$

iii.   What is the missing number: $(49 − 63 + 91) =$ **?**$(7 − 9 + 13)$

iv. $|\text{-}24| \div 2(5 − 1) − 16 \div 8 =$

v. $8(14 − 6) \div |4 − 8| =$

vi. $8(14 − 6) + 3(|\text{-}7| − 3) =$

vii. $2(17 − 3 − 4 + 9) − 8(12 \div 4 \times 2) =$

Chapter 7 – Introduction to Variables

7.1 Our Focus

In mathematics we work with unknown values. We assign letter names to unknowns and refer to them as **variables**. In this chapter, we will familiarize ourselves with basic operations involving variables, solve simple variable problems, and work with simple variable expressions. In a subsequent chapter, we will delve more deeply into variables.

7.2 Basics of Operations Involving Variables

A **variable** represents a number or expression whose actual value is not fixed or known. Sometimes we use a variable as a place holder for any number we can think of. Other times, the facts of the problem we are faced with define the value, or range of values, a variable can have. In order to use a variable in an expression, we assign it an identifier. The identifier can be any letter of the alphabet.

In a problem involving a variable, we need enough information to determine the variable's actual value.

We will start our discussion with three variables: **a**, **b**, and **c.** What we say about them applies to all variables.

We can add **a** and **b** just like we add **2** and **3**. We write: **a + b**.

Since we have not yet assigned numeric values to **a** and **b**, we can't evaluate the expression **a + b** to get a numeric result.

We can subtract **a** from **b**. We write: **b – a**.

We can multiply **a** and **b**. We write: **a × b** or **b × a**.

We usually drop the multiplication symbol and write: **ab** or **ba**.

We can multiply variables and numerals. When we multiply the number **2** and the variable **a**, we write: **2a**. We write the numeral first. When we multiply the number **2** and the variables **a** and **b**, we write **2ab**.

Obviously, this way of writing multiplication does not apply to numerals; we do not write **2 × 3** as **23** because the two forms have very different meanings.

When we **multiply** a variable by **1** the result is the variable: **a × 1 = a**. The multiplication identity applies to variables.

When we **multiply** a variable by zero, the result is zero: **a × 0 = 0**.

We can divide **b** by **a**, and vice versa. We write **b** divided by **a** as:

$$\mathbf{b \div a} \quad \text{or} \quad \mathbf{{}^{b}/_{a}}.$$

We write **a** divided by **b** as:

$$\mathbf{a \div b} \quad \text{or} \quad \mathbf{{}^{a}/_{b}}.$$

We can divide variables by numerals and vice versa. When we divide **a** by **5** we write:

$a \div 5$ or $^a/_5$,

and when we divide **5** by **a** we write:

$5 \div a$ or $^5/_a$.

When we **divide** a variable by **1** the result is the variable: $a \div 1 = a$. We cannot divide anything by zero because the result is undefined.

When we add **b** to itself **3** times we get: $b + b + b = 3b$, which is equivalent to: $b \times 3 = 3 \times b = 3b$.

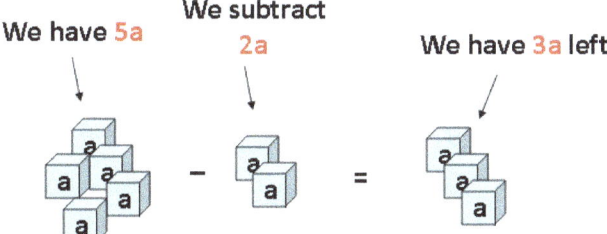

We illustrate this in the graphic above. Think of the variable **b** as a basket containing an unknown number of oranges. We can add **3** such baskets to get (**3b**), or 3 baskets of oranges, *not just the numeral 3*.

Similarly, if we have **3** such baskets (**3b**), we can take away or subtract one basket (**b**), and be left with **2** baskets (**2b**), *not just the numeral 2*. We can do all this without knowing the number of oranges in a basket. We illustrate with the next graphic.

Similarly, subtracting **2a** from **5a** gives **3a**, as shown in the next graphic.

Because we don't know how many oranges are in a basket, if we stick our hands into a basket and remove **3** oranges, we can't say how many oranges are left in the basket. We will not know this until we discover the quantity of oranges originally in the basket. This is illustrated in the graphic below.

b and **3** are factors of **3b** because we multiply **b** and **3** to get **3b**.

If, for example, we determine that **b = 12**, then (**b − 3**) becomes: **12 − 3 = 9**.

Dividing **5a** by **5** gives **a**: $\frac{5a}{5} = a$; we reduce by the common factor **5**.

and dividing **5a** by **a** gives **5**: $\frac{5a}{a} = 5$; we reduce by the common factor **a**.

If, for example, we assign the value **4** to **a**, **5a** becomes **20**, and **5a ÷ a** becomes **20 ÷ 4**, which gives **5**.

Everything we have said so far is consistent with what we learned about operations involving numerals. It further confirms that variables are just unknown values. There is nothing mysterious about them!

7.3 Determining the Value of a Variable - Basics

Let us examine the following statement:

> *Example 1*: 2 + **b** = 5.

What is the numeric value of **b**?

In this example **b** is a variable. It represents an unknown value. In this case, the statement gives us enough information to determine that **b** = 3, because **3** is the number we add to **2** to get **5**. The following graphic illustrates how we work out the value of **b** through a simple mental exercise.

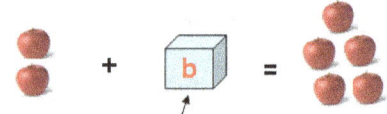

There must be 3 apples in here!

b = 3 apples

Let us look at another example:

> *Example 2*: **b** × 3 = 12; **b** = ?

Again, we are working with a variable named **b**. The statement says that multiplying the variable **b** by **3** gives **12**; what is the numeric value of **b**? From our knowledge of multiplication, **4 × 3 = 12**, therefore, we conclude that **b** = 4. The graphic below illustrates this.

b × 3   =   12

There is got to be 4 apples in each box!

Let us look at another example.

Example 3: **12 ÷ b = 4; b = ?**

Once more we are working with a variable named **b**. The statement says that dividing **12** by the variable **b** gives **4**; what is the numeric value of **b**? From our knowledge of division:

> **12 ÷ 3 = 4**

Therefore, we conclude that **b** = **3**.

Examples 2 and *3* give different information about **b**. As a result, the value we determine for **b** in each case is different.

7.4 Simplifying a Variable Expression - Basics

The purpose of simplifying a variable expression is to make the expression simpler, clearer, and easier to evaluate. We use the order of operation rule to simplify a variable expression. Let us work on the following example which involves just one variable: **a**.

> *Example 1*: 5a + 13a − 6a ÷ 3
>
> Step 1: perform **6a ÷ 3**; the result is **2a**.
>
> Step 2: perform **5a + 13a − 2a** = **16a**

 We did not evaluate the expression in *Example 1* because we do not know the numeric value of the variable **a**. We only simplified the expression to make it less complicated looking. If **a** is assigned a value, **16a** will obviously be easier to evaluate than the original expression.

We will discuss a few new topics then pick up variables again. This will allow us to give variables a full treatment.

7.5 Review Problems

i.   x + 12 = 14; x = _____

ii.   b − 13 = 17; b = _____

iii. 3b − 2 = 16; b = _____

iv. 3x ÷ 4 = 36; x = _____

v.   Simplify: 15a ÷ 3 − 6a ÷ 3 = _____

vi. Evaluate the following expressions, given **a** = 2, **b** = 5, **x** = 3,

   a)   2b − 2a + 7x

   b)   35x ÷ 7 − 5x + 9a

Chapter 8 – Factoring and Fractions

8.1 Our Focus

We intend to discuss fractions, but before we do that, we will have a brief discussion of factors and factoring. This is necessary because we often have to reduce a fraction to its lowest terms. The process of reducing a fraction involves factoring. Also, an understanding of factors is required in operations involving fractions. Once we examine factors and factoring, we will move on to basic operations involving fractions. Then we will move to more complex expressions, including expressions with fractions and integers. We will also discuss fractions consisting of variables.

8.2 Factors and Factoring

If **a**, **b**, and **c** are integers and **a** × **b** = **c**, then **a** and **b** are factors of **c**. To clearly understand what this means, we will use real integers in place of **c**, then find the factors that **a** and **b** represent. We will let **c** equal **8**.

Example 1: What are the factors of **8**.

From the multiplication table:

$$2 \times 4 = 8, \text{ and}$$

$$1 \times 8 = 8$$

The factors of 8 are: **1, 2, 4**, **8**; we can multiply these numbers in various pairs to get **8**.

Any integer is a factor of itself, and the number **1** is a factor of every integer.

A number is **divisible** by its factors. This makes sense, since we multiply the factors to get the number. Again, *divisible* means *divides without a remainder*. From ***Example 1***:

$$8 \div 2 = 4; \text{ no remainder.}$$

$$8 \div 4 = 2; \text{ no remainder.}$$

$$8 \div 1 = 8; \text{ no remainder.}$$

$$8 \div 8 = 1; \text{ no remainder.}$$

Compare these results to:

$$8 \div 3 = 2^2/_3.$$

Dividing **8** by **3** does not give an integer result. Thus, **3** is not a factor of **8**.

Example 2: What are the factors of 30?

From the multiplication table:

$$2 \times 15 = 30,$$

3 × 10 = 30, and

5 × 6 = 30.

Also, **1** and **30** are factors of **30**.

Therefore, the factors of 30 are: **1**, **2**, **3**, **5**, **6**, **10**, **15**, **30**.

Let us look at another example:

    Example 3: What are the factors of 24?

From our discussion above, we know immediately that **1** and **24** are among the factors of **24**.

From our knowledge of multiplication, we know that:

    8 × 3 = 24.

Therefore, **8** and **3** are also factors of 24.

Because **8** is a factor of **24**, the other factors of **8**: **2** and **4**, are also factors of **24**.

So far, we have the following factors: **1**, **2**, **3**, **4**, **8**, **24**.

Factors come in pairs. The factor **2** pairs with **12**, and the factor **4** pairs with **6**, because: **2** × **12** = **24**, and **4** × **6** = **24**, making **12** and **6** also factors of **24**.

The complete list of factors of **24** is: **1**, **2**, **3**, **4**, **6**, **8**, **12**, **24**.

To make sure our list is complete, we go through it and verify that each factor we have identified has a pair that multiplies it to give **24**.

We make extensive use of factoring in all the topics that follow, particularly in fractions. Thus knowledge of multiplication is critical from this point forward.

8.3 Introduction to Fractions

A fraction is a number that has been expressed as a ratio of two integers. The number $^1/_2$ is a fraction. It has been expressed as the ratio of the integers **1** and **2**. In the fraction form, the number on top is called the **numerator**; the number on the bottom is called the **denominator**.

The number $^1/_2$ has a specific meaning. The graphic below depicts $^1/_2$.

1 apple   2 half apples

When we divide **1** by **2**, the result is $^1/_2$ in fraction form. When we divide **1** by **10**, the result is $^1/_{10}$ in fraction form. If we were to use a graphic like the one above to illustrate $^1/_{10}$, we would have **10** equal pieces of one apple.

What happens when we divide **2** by **4**? Again we use a graphic to illustrate the result.

2 apples   **4 half apples**

   ÷ 4 =   = $^1/_2$

We write **2** divided by **4** as: $^2/_4$ in fraction form. As the graphic shows, $^2/_4$ equals $^1/_2$. How so? Because the ratio: **two-to-four** is equivalent to the ratio: **one-to-two**. We can divide two-to-four into two groups of one-to-two. The ratio remains unchanged. You can see this from the graphic above.

We write **3** divided by **15** as $^3/_{15}$ in fraction form, **10** divided by **3** as $^{10}/_3$ in fraction form, **45** divided by **50** as $^{45}/_{50}$ in fraction form, and so on.

8.4 Reducing A Fraction

We said above that:

$$^2/_4 = {}^1/_2$$

We usually reduce a fraction until its numerator and denominator have no common factors. This maintains the ratio, while bringing the fraction to its lowest terms. The fraction $^2/_4$ reduces to $^1/_2$, so we would write $^1/_2$ for a result instead of $^2/_4$.

How do we reduce a fraction? We write both the numerator and the denominator as products of their relevant factors. Any factor that appears in the numerator and has a pair in the denominator is replaced with the number **1**. Once we have completed the replacements, we multiply the numbers in the numerator to get its reduced value. We similarly multiply the numbers in the denominator to get its reduced value.

Let us look at some examples.

Example 1: How do we reduce $^2/_4$ to $^1/_2$?

We write $^2/_4$ as: $^2/_{2\times2}$.

We replace **2** in the numerator and a matching **2** in the denominator each with **1**. We get: $^1/_{1\times2}$

We are left with **1** in the numerator. We multiply the numbers left in the denominator to get **2**. The reduced fraction is $^1/_2$.

Example 2: How do we reduce $^3/_{15}$?

We write $^3/_{15}$ as: $^3/_{3\times5}$. Notice that we did not bother to write the factor **1** because it really does not affect the eventual multiplication.

We replace the factor **3** with **1** in both appearances. We get: $^1/_{1\times5}$, and the final result is $^1/_5$.

Example 3: How do we reduce $^{50}/_{100}$?

$$^{50}/_{100} => {}^{50}/_{50\times2} => {}^1/_{1\times2} = {}^1/_2$$

Example 4: How do we reduce $^{7a}/_{21a}$?

The factors of **7a** are **7** and **a**. The relevant factors of **21a** are **3**, **7** and **a**. We can see that **7** and **a** are factors of both the numerator and the denominator. We replace both appearances of **7** with **1**, and we likewise replace both appearances of **a** with **1**.

We get: $^{7a}/_{21a} => {}^{7 \times a}/_{3 \times 7 \times a} = {}^{1 \times 1}/_{3 \times 1 \times 1} = {}^{1}/_{3}$

Note that we handled the variable **a** the same way we handled the numeric factor **7**; we replaced **a** with **1** because it appears in both the numerator and the denominator.

Example 5: How do we reduce $^{12}/_{14}$?

$^{12}/_{14} => {}^{2 \times 6}/_{2 \times 7}$

We replace the **2**s with **1**. The next step becomes:

$^{12}/_{14} => {}^{1 \times 6}/_{1 \times 7} => {}^{6}/_{7}$

The result tells us that the numbers 6 and 7 have no factors in common and we can't reduce the fraction $^{6}/_{7}$ any further.

Example 6: How do we reduce $^{63}/_{14}$?

$^{63}/_{14} => {}^{9 \times 7}/_{2 \times 7}$

In this example, 7 is a common factor, so we replace it with the number **1** in both places, and get:

$^{9 \times 1}/_{2 \times 1} => {}^{9}/_{2} => 4^{1}/_{2}$

When the numerator of a fraction is larger than the denominator, it is called an **improper fraction**. Thus, $^{9}/_{2}$ is an improper fraction.

An improper fraction may be converted to a mixed number. We learned earlier that a mixed is made up of an **integer part** and a **fraction part**. Our result: $4^{1}/_{2}$ is a mixed number. To get it, we divide **9** by **2** to get **4**, with a remainder of **1**. The fraction part $^{1}/_{2}$ is formed by the remainder **1** and the divisor **2**.

When we perform operations involving fractions and mixed numbers, we often convert the mixed numbers to improper fractions.

To convert a mixed number to an improper fraction we follow the following steps:

Step 1: We multiply the whole number with the denominator of the fraction to get a product. Call the product **X**.

Step 2: We add **X** and the numerator of the fraction to get a sum. Call the sum **Y**.

Step 3: We make **Y** the numerator of the improper fraction.

Step4: We make the denominator of the fraction the denominator of the improper fraction.

For example, to go from $4^{1}/_{2}$ to the improper fraction $^{9}/_{2}$, we multiply **4** and **2** to get **8**. Then we add **1** (from the fraction numerator) to **8** to get **9**, which becomes the numerator of the improper fraction. **2** remains the denominator.

8.5 Even, Odd, Prime Numbers and Fraction Reduction

Let us see how even, odd, prime, and other properties of numbers help us in fraction reduction.

A fraction consisting of a prime number and a non-prime cannot be reduced unless the non-prime is a multiple of the prime number. For example, we cannot reduce $^{11}/_{32}$, but we can reduce $^{11}/_{33}$ to $^{1}/_{3}$ because **33** is a multiple (by **3**) of **11**.

If both the numerator and the denominator of a fraction are even, we know right away that we can do at least one initial reduction by dividing each of them by **2**. We continue reducing this way until one of the numbers is no longer even. At that point we test for possible common odd factors like **3**, **5**, **7**, **9**, **11**, **13**, and so on. If we determine that the two numbers no longer have common factors, we have accomplished our mission.

Familiarity with multiplication allows us to make some observations about whole numbers and their factors. We already know that **even** numbers are divisible by **2**. Whole numbers that end in **5** are divisible by **5**. Whole numbers that end in zero (**0**) are divisible by both **5** and **10**. We also know that if a whole number ends in **1** or **3** or **9** it may be divisible by either **3**, **7**, **9**, **11**, or **13**. This knowledge can simplify the task of reducing a fraction.

Let us look at one more example of reducing a fraction, with these observations in mind:

Example 1: $^{616}/_{968}$

On the surface, this looks like a daunting task, but observation tells us that both the numerator and the denominator are even. Dividing each number by **2**, we arrive at:

$^{308}/_{484}$

We again have two even numbers, so we again divide by 2 and get:

$^{154}/_{242}$

Another iteration of dividing by 2 gives us:

$^{77}/_{121}$

From our knowledge of multiplication, we can tell that 77 is **11 × 7**, and 121 is **11 × 11**. So, **11** is a common factor. Dividing both **77** and **121** by **11**, gives us the final result: $^{7}/_{11}$. This is the same as replacing **11** in both the numerator and the denominator by **1**.

By using our knowledge of this property of even numbers, we have been able to reduce a fraction with big numbers to its lowest term without much mental strain!

8.6 Review Problems

Without using a calculator, reduce the following fractions. Convert improper fractions to mixed numbers.

i. $^{13}/_{26}$

ii. $^{96}/_{36}$

iii. $^{21}/_{147}$

iv. $^{85}/_{95}$

v. $^{90}/_{1080}$

vi. $^{288}/_{1440}$

vii.   What are the factors of the number **54**?

viii.   Identify the prime numbers in the following list: 2, 16, 29, 33, 61, 71

ix.   A boy has 3 boxes of legos. The boxes are labeled A, B, and C. Box A contains 81 pieces. Box B contains 27 pieces. Box C contains 54 pieces. What fraction of the total number of pieces does box C have? Reduce your answer to the lowest terms.

8.7 Fractions – Addition and Subtraction

We have already looked at the structure of fractions. We said the number on top is called **numerator**, and the number on the bottom **denominator**. Now we will look at arithmetic operations involving fractions. Let us start with a very simple example:

> ***Example 1***: $^{1}/_{2} + ^{1}/_{2}$

In this example, both fractions have the same denominator. When we add fractions that have a **common denominator**, we retain the denominator and simply add the numerators. In this example, we keep the denominator **2** and add **1** and **1**. The result is: $^{2}/_{2}$, which equals **1**. The graphic below confirms the result.

½ an apple   **½ an apple**   **1 apple**

+   =

Another example:

> ***Example 2***: $^{1}/_{6} + ^{1}/_{6}$

Again, the fractions have a common denominator. We simply add the numerators and retain the denominator. In this example, we keep the denominator **6** and add **1** and **1**. The result is: $^{2}/_{6}$.

Since $^{2}/_{6}$ is not in reduced form, we reduce it to: $^{1}/_{3}$.

Another example:

> ***Example 3***: $^{5}/_{7} + ^{1}/_{7} - ^{2}/_{7}$

Again, we have a common denominator: **7**. We retain **7** as the denominator of the result, and evaluate the expression using the numerators:

$$5 + 1 - 2 = 4$$

The final result is: $^4/_7$.

Another example:

Example 4: $^1/_2 - ^2/_3 + ^3/_5$

In this case, the three fractions have different denominators. In order to successfully evaluate the expression, we must bring the fractions to a common denominator without changing the ratios that the fractions represent.

The common denominator we want is the **smallest** number for which **2**, **3**, and **5** are factors. That number is called **the least common denominator**. Here we are working backwards, the opposite of what we previously did with factoring. In factoring, we started with a big number and broke it down into its factors. Here, we have factors and we want the smallest number that the factors all belong to. The smallest number that has **2**, **3**, and **5** as factors is **30**. We get **30** by multiplying the three numbers: **2**, **3**, **5**.

Now that we have the least common denominator, we have to modify each fraction so that its denominator becomes **30**, while it maintains its original ratio.

The first fraction: $^1/_2$ becomes: $^{15}/_{30}$.

Notice that the ratio has remained the same. We get $^{15}/_{30}$ in these steps:

- Divide 30 by the denominator 2 to get 15
- Multiply 15 by the numerator 1 to get 15
- Make 30 the new denominator and 15 the new numerator.
- The result is: $^{15}/_{30}$.

The next fraction: $^2/_3$ becomes: $^{20}/_{30}$.

Again, the ratio has remained the same. We get $^{20}/_{30}$ in these steps:

- Divide 30 by the denominator 3 to get 10.
- Multiply 10 by the numerator 2 to get 20
- Make 30 the new denominator and 20 the new numerator.
- The result is: $^{20}/_{30}$

The final fraction: $^3/_5$ becomes: $^{18}/_{30}$.

Again, the ratio has remained the same. We get $^{18}/_{30}$ in these steps:

- Divide 30 by the denominator 5 to get 6.
- Multiply 6 by the numerator 3 to get 18
- Make 30 the new denominator and 18 the new numerator.
- The result is: $^{18}/_{30}$.

Now, we re-write the original expression using the new fractions we have created.

$^1/_2 - ^2/_3 + ^3/_5$ becomes $^{15}/_{30} - ^{20}/_{30} + ^{18}/_{30}$.

Note that we have not changed the original expression. We have simply converted the fractions to **unreduced** forms that have *a common denominator*. After we evaluate the expression, we reduce the final result, if necessary.

Now that the fractions have a common denominator, all we have to do is evaluate the expression using the numerators:

$^{(15 - 20 + 18)}/_{30}$

$= ^{13}/_{30}$

Since **13** and **30** have no common factors, we have our final result: $^{13}/_{30}$.

Another example:

Example 5: $^1/_2 - ^1/_4 + ^3/_5$

Again, we need to find the least common denominator. We notice that **2** is a factor of **4**. It means that the number that has **4** and **5** as factors will automatically have **2** as a factor. So we multiply only **4** and **5**. The least common denominator is **20**.

Now we follow the steps we laid out above to convert the fractions to a common denominator.

$^1/_2$ becomes $^{10}/_{20}$

$^1/_4$ becomes $^5/_{20}$

$^3/_5$ becomes $^{12}/_{20}$

The original problem: $^1/_2 - ^1/_4 + ^3/_5$ becomes: $^{10}/_{20} - ^5/_{20} + ^{12}/_{20,}$

and the result is: $^{17}/_{20}$

Another example:

Example 6: $^a/_3 + ^a/_5$

The two fractions have denominators **3** and **5**. The least common denominator is **15**. Following our steps,

$^a/_3$ becomes $^{5a}/_{15}$

$^a/_5$ becomes $^{3a}/_{15}$

The original problem $^a/_3 + ^a/_5$ becomes:

$^{5a}/_{15} + ^{3a}/_{15} = ^{8a}/_{15}$

Since **8a** and **15** have no common factors, $^{8a}/_{15}$ is the final result.

It is important to note that the presence of the variable **a** did not change our approach. As you know, we perform operations with variables in the same way we do numerals.

8.7.1 Review Problems

Without using a calculator, solve the following problems. Convert results that are improper fractions to mixed numbers.

i. $^1/_2 + {}^2/_5 =$

ii. $^1/_8 + {}^7/_9 =$

iii. $^{11a}/_{12} - {}^a/_6 =$

iv. $^{7x}/_8 - {}^x/_8 =$

v. $^1/_3 + {}^5/_6 - {}^2/_5 + {}^4/_3 =$

vi. $^1/_2 + {}^1/_3 + {}^1/_4 + {}^1/_5 =$

vii.   One brother brought half a cup of cereal to the breakfast table. The second brother brought a third of a cup. The third brother brought a quarter of a cup. What fraction of a cup do the three portions add up to?

8.8 Fractions - Multiplication and Division

In the previous discussion, we learned the steps for performing fraction addition and subtraction. The steps for fraction multiplication and division are different from the steps for fraction addition and subtraction. For example, we do not find a least common denominator. In this sense, the process for multiplying and dividing fractions is more straight forward and simpler.

We will first discuss fraction multiplication. Then we will examine the relationship between fraction multiplication and fraction division, and how to change a fraction division problem to a fraction multiplication problem.

8.8.1 Fractions - Multiplication

Multiplying fractions is simple. We multiply the numerators of the fractions to get the numerator of the result. Then we multiply the denominators of the fractions to get the denominator of the result. We reduce the result as necessary. The graphic below illustrates fraction multiplication.

Multiply the numerators   **Reduce the result**

$$\frac{2}{3} \times \frac{3}{7} = \frac{6}{21} = \frac{2}{7}$$

Multiply the denominators

Let us work through some examples.

Example 1: $^5/_7 \times {}^{11}/_{17}$

1. Multiply 5 and 11 to get 55.
2. Next, multiply 7 and 17 to get 119.
3. The result is $^{55}/_{119}$. This fraction cannot be reduced any further because 55 and 119 do not have common factors. We are done!

Remember that the numerators **5** and **11** are factors of their product **55**. And the denominators **7** and **17** are factors of their product **119**. By examining the factors, we quickly determine that $^{55}/_{119}$ cannot be reduced.

In fraction multiplication, it is sometimes beneficial to reduce the fractions **prior to** performing the actual fraction multiplication. To do this, we replace a factor that appears in the numerator of any of the fractions **and also appears** in the denominator of any of the fractions with **1** in both places. We repeat the process until we no longer have a factor that is common to a numerator and denominator. Now we multiply numerators to get the numerator of the result and denominators to get the denominator of the result. Reducing ahead lets us multiply smaller numbers and arrive at an already reduced final result.

Let us look at an example. First, we will multiply without reducing first, then we will reduce prior to multiplying.

Example 2: $^{2a}/_3 \times {}^3/_5$

Following the same steps that we used in the previous example, we have:

2a × 3 = 6a (numerator of result)

3 × 5 = 15 (denominator of result)

The result of multiplying the two fractions is $^{6a}/_{15}$. We notice that **3** is a common factor of both **6a** and **15**, so we reduce $^{6a}/_{15}$ to get the final result: $^{2a}/_5$.

Now, let us reduce before the multiplication. Here is the original expression:

$^{2a}/_3 \times {}^3/_5$

We see that the factor **3** appears in the numerator of one fraction and in the denominator of another fraction. We replace it with **1** in both places. We get:

$= {}^{2a}/_1 \times {}^1/_5 = {}^{(2a \times 1)}/_{(1 \times 5)}$

$= {}^{2a}/_5.$

We get a final result that is already reduced!

Why does reducing ahead work? To answer this question, let us go back to **Example 2**:

$^{2a}/_3 \times {}^3/_5.$

From the commutative property, the product of the numerators: (2a × 3) can be re-written as: (3 × 2a). Therefore, the fraction multiplication problem can also be re-written as:

$^3/_3 \times {}^{2a}/_5$.

Now, since any number divided by itself equals **1**, $^3/_3$ reduces to **1**.

$^3/_3 \times {}^{2a}/_5 = \mathbf{1} \times {}^{2a}/_5 = {}^{2a}/_5$.

When we reduce ahead of fraction multiplication, we do not need to re-arrange numerators like we did above to answer the question. We accept the underlying idea and move forward to save time and effort.

8.8.2 Fractions - Division

Before we delve into fraction division, let us define the **reciprocal** of a fraction.

We get the reciprocal of a fraction by **inverting** the fraction so that its denominator becomes the numerator, and its numerator becomes the denominator. This is shown in the graphic below.

Original fraction   Its reciprocal

$^3/_5$ => $^5/_3$

Needless to say, the two fractions, $^3/_5$ and $^5/_3$, are reciprocals of each other.

We also get the reciprocal of a fraction when we divide the number **1** by the fraction. For example, when we divide **1** by $^2/_3$ the result is $^3/_2$, as shown below.

$$1 \div \frac{2}{3} = \frac{1}{\frac{2}{3}} = \frac{3}{2}$$

This tells us that the product of a fraction and its reciprocal equals **1**.

Now to fraction division. The process for dividing one fraction by another differs from fraction multiplication only in the first two steps.

Step 1: We convert the divisor to its **reciprocal**.

Step 2: We change the division operator to the multiplication operator.

Step 3: We perform fraction multiplication as usual.

The pictorial that follows illustrates the process of fraction division. Of course, we would convert the final result $^{14}/_9$ to the mixed number: $1^5/_9$.

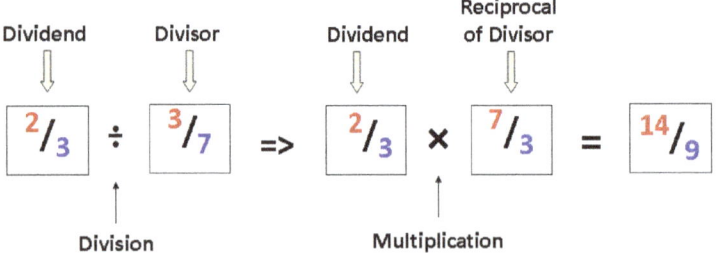

Let us work through an example.

Example 1: $^2/_3 \div {}^3/_5$

In this example, $^2/_3$ is the dividend, $^3/_5$ is the divisor. The divisor is the fraction to the right of the division operator. We replace $^3/_5$ with its reciprocal $^5/_3$, and change division to multiplication.

We go **from:** $^2/_3 \div {}^3/_5$ **to:** $^2/_3 \times {}^5/_3$

Now we can apply what we have already learned about fraction multiplication to evaluate the expression. The result is $^{10}/_9$ or $\mathbf{1^1/_9}$.

Why does using the reciprocal and changing to multiplication work?

Let us see what dividing one number by another number really entails. We will use **a** and **b** to stand for any two numbers.

$a \div b = {}^a/_b$, and $^a/_b = a \times {}^1/_b$

Therefore:

$a \div b = a \times {}^1/_b$.

Now, let us replace **a** and **b** with the fractions: $a = {}^1/_2$, $b = {}^2/_3$. Then,

$$a \div b = \quad \cfrac{\frac{1}{2}}{\frac{2}{3}}$$

But, as we learned earlier,

$$\cfrac{\frac{1}{2}}{\frac{2}{3}} = \frac{3}{2}$$

Therefore:

$$a \div b = \frac{1}{2} \times \frac{3}{2} = \frac{3}{4}$$

This is the same result as taking the reciprocal of $^2/_3$ and changing division to multiplication.

8.8.3 Review Problems

Without using a calculator, solve the following problems. Reduce the result, if it is not in reduced form.

i. $^4/_5 \times {}^5/_7 =$

ii. $^a/_{11} \div {}^{7a}/_{77} =$

iii. $^2/_3 \times {}^7/_8 =$

iv. $\quad {}^{2x}/_3 \times {}^1/_{3x} =$

v. $\quad {}^{12}/_{25} \times {}^5/_9 =$

vi. $\quad {}^3/_{7x} \div {}^5/_{7x} =$

vii. A cook poured half of the oil in a bottle into a cup. He then poured a third of the oil in the cup into a frying pan. What fraction of bottle's original content ended up in the frying pan?

8.9 Mixed Operations with Fractions

Now that we have a good handle on fractions, let us look at more complex expressions involving fractions.

Example 1: $\quad {}^1/_2 - {}^1/_3 \times {}^3/_5 \div {}^1/_7 + {}^2/_9$

We use the same rules we have learned about evaluating arithmetic expressions and handling fractions, no matter how complex the expression. We follow the order of operation rule. When a fraction is a divisor, we take its reciprocal and change the operation to multiplication. We determine least common denominator for addition and subtraction operations as necessary. And so on.

Having reviewed our rules, let us work on **Example 1**. First we perform multiplication and division:

- We replace ${}^1/_7$ with its reciprocal ${}^7/_1$, and change the division to multiplication:

  ${}^1/_3 \times {}^3/_5 \div {}^1/_7$ becomes ${}^1/_3 \times {}^3/_5 \times {}^7/_1$.

- Now we multiply the numerators: $1 \times 3 \times 7 = 21$, and multiply the denominators: $3 \times 5 \times 1 = 15$. From this step we have:

  ${}^1/_3 \times {}^3/_5 \times {}^7/_1 = {}^{21}/_{15}$,

- We reduce ${}^{21}/_{15}$ to ${}^7/_5$, since 3 is a common factor of 21 and 15. (Note that we could have first reduced: ${}^1/_3 \times {}^3/_5 \times {}^7/_1$ **to:** ${}^1/_1 \times {}^1/_5 \times {}^7/_1$ and immediately arrived at the final result ${}^7/_5$.)

- We substitute the intermediate result ${}^7/_5$, and the original expression goes from:

  ${}^1/_2 - {}^1/_3 \times {}^3/_5 \div {}^1/_7 + {}^2/_9$ **to:** ${}^1/_2 - {}^7/_5 + {}^2/_9$

- Now we evaluate: ${}^1/_2 - {}^7/_5 + {}^2/_9$.

  The least common denominator is 90, so the expression becomes:

  ${}^{45}/_{90} - {}^{126}/_{90} + {}^{20}/_{90}$

  The denominator of the result is (90). Evaluating using the numerators:

  $45 - 126 + 20$

  $= -81 + 20 = -61.$

The final result is: $\left(-\frac{61}{90}\right)$.

Below, we have modified the same expression to have a grouped part. Let us see if the final result differs.

Example 2: $\left(\frac{1}{2} - \frac{1}{3}\right) \times \frac{3}{5} \div \frac{1}{7} + \frac{2}{9}$

- We do the grouped part first:

$$\frac{1}{2} - \frac{1}{3} = \frac{3}{6} - \frac{2}{6} = \frac{1}{6}$$

- The original expression becomes:

$$\frac{1}{6} \times \frac{3}{5} \div \frac{1}{7} + \frac{2}{9}$$

- We do the multiplication and division next:

$$\frac{1}{6} \times \frac{3}{5} \div \frac{1}{7}$$
$$= \frac{1}{6} \times \frac{3}{5} \times \frac{7}{1}$$
$$= \frac{21}{30} = \frac{7}{10}$$

- Finally, we have:

$$\frac{7}{10} + \frac{2}{9}$$
$$= \frac{63}{90} + \frac{20}{90}$$
$$= \frac{83}{90}, \text{ which is the final result.}$$

Clearly, having to do the grouped addition first made a difference in the result.

8.10 Expressions Involving Fractions and Integers

Evaluating an expression consisting of fractions and whole numbers is easy. Every whole number can be written as a fraction with a denominator of **1**. We convert the whole number in the expression to fraction form with denominator 1. For example, **2** becomes $\frac{2}{1}$. Recall that when we divide a number by **1** we get the original number. So, writing **2** as $\frac{2}{1}$ does not change its value.

If the expression contains a mixed number, we convert the mixed number to an improper fraction. Once every term of the expression is in fraction form, we simply apply our rules for evaluating a fraction expression.

With this in mind, let us evaluate the following expression:

Example 1: $\frac{1}{2} - \frac{1}{3} \times \frac{1}{7} \div 2$.

- We re-write the expression as: $\frac{1}{2} - \frac{1}{3} \times \frac{1}{7} \div \frac{2}{1}$.
- We perform multiplication and division:

$$\frac{1}{3} \times \frac{1}{7} \div \frac{2}{1}$$

$$= \frac{1}{3} \times \frac{1}{7} \times \frac{1}{2} = \frac{1}{42}.$$

- We substitute this intermediate result and perform the subtraction:

$$\frac{1}{2} - \frac{1}{42}$$

$$= \frac{21}{42} - \frac{1}{42}$$

$$= \frac{20}{42} = \frac{10}{21}. \quad \text{(Task accomplished!)}$$

8.11 Fractions made up of Variables

We will use several examples to discuss how we handle expressions that include fractions made up of variables. Let us start with the following example:

Example 1: $\frac{2}{a} + \frac{1}{a}$

As we learned in our earlier discussion of fraction addition, when we add two fractions that have the same denominator, we simply add the numerators. Thus, the result of **Example 1** is: $\frac{3}{a}$. It does not matter that the denominators are not numerals.

Let us look at another example:

Example 2: $\frac{2}{a} + \frac{1}{5}$

This time the fractions have different denominators: the variable **a** and the numeral **5**. Their product is **5a**. We make this the least common denominator. Now we convert the two fractions so that they both have the denominator **5a**.

For the fraction: $\frac{2}{a}$

- Divide **5a** by the fraction's denominator **a** to get **5**.
- Multiply **5** by the fraction's numerator **2** to get **10**.
- Make **5a** the new denominator and **10** the new numerator.
- The fraction $\frac{2}{a}$ becomes: $\frac{10}{5a}$.

For the fraction: $\frac{1}{5}$

- Divide **5a** by the fraction's denominator **5** to get **a**.
- Multiply **a** by the fraction's numerator **1** to get **a**.
- Make **5a** the new denominator and **a** the new numerator.
- The fraction $\frac{1}{5}$ becomes: $\frac{a}{5a}$.

Now that the fractions have a common denominator, we proceed with the addition. The result is:

$$\frac{10 + a}{5a}$$

Let us look at another example.

Example 3: $^2/_a + ^3/_b$

Again, the fractions have different denominators. We make their product **ab** the least common denominator. Now, we convert the two fractions so that they both have the denominator **ab**.

For the fraction: $^2/_a$

- Divide **ab** by the fraction's denominator **a** to get **b**.
- Multiply **b** by the fraction's numerator **2** to get **2b**.
- Make **ab** the new denominator and **2b** the new numerator.
- The fraction $^2/_a$ becomes: $^{2b}/_{ab}$

For the fraction: $^3/_b$

- Divide **ab** by the fraction's denominator **b** to get **a**.
- Multiply **a** by the fraction's numerator **3** to get **3a**.
- Make **ab** the new denominator and **3a** the new numerator.
- The fraction $^3/_b$ becomes: $^{3a}/_{ab}$

Now that the fractions have a common denominator, we proceed with the addition. The result is:

$$\frac{2b + 3a}{ab}$$

8.12 Review Problems

Without using a calculator, solve the following problems. Reduce the result, if necessary.

i. $1 + {}^1/_2 + {}^1/_3 + {}^1/_4 + {}^1/_5 =$

ii. ${}^1/_{13} \times {}^{39}/_7 \div {}^5/_7 + 1{}^1/_3 \times {}^1/_5 =$

iii. ${}^9/_{16} - {}^1/_4 + 1{}^5/_8 =$

iv. $1 - {}^1/_{2a} - {}^1/_{3b} =$

v. ${}^1/_{ac} + {}^5/_{7b} =$

vi. $2{}^5/_6 + {}^{39}/_7 \div {}^{13}/_7 \times ({}^1/_5 - {}^1/_6) =$

vii.   Two brothers went fishing and caught a lot of fish. They gave $^1/_3$ of their catch away to friends who did not catch any. Then they gave $^1/_2$ of what was left to their neighbors. They were left with 12 fish. If one boy caught twice as many fish as the other, including what they gave away, how many fish did each boy catch?

Chapter 9 - Decimals

9.1 Our Focus

A decimal is simply another form for a fraction. In this chapter, we will first examine the structure of a decimal number. Then, we will discuss how to convert a fraction to decimal form, and how to perform arithmetic operations with decimals.

9.2 The Structure of a Decimal

The decimal form is simply another way we represent a fraction or a mixed number. For example, we can write the result of the division **23 ÷ 4** as the mixed number $5^3/_4$, or in decimal form: **5.75**. In mathematics, what form we choose is dictated by the situation.

In decimal form, we write the whole number part and place a dot after it. Then we write the fraction part as multiples of $^1/_{10}$, $^1/_{100}$, $^1/_{1000}$, ..., and so on. The dot after the whole number is called the **decimal point**.

The $dollar currency, which we are quite familiar with, provides an excellent basis for understanding decimals because it is based on the decimal system. A penny represents $^1/_{100}$th of a dollar. Ten pennies equal a dime, which represents $^1/_{10}$th of a dollar. Ten dimes add to a single dollar. We can exchange ten singles for one ten-dollar note. We can exchange ten $10 bills for a hundred-dollar bill, and so on. A value of **1** in a position represents $^1/_{10}$th of the value of the denomination immediately to its left.

In the graphic below we have broken down the dollar amount **$238.75** into its parts to illustrate the concept of a decimal number. The dimes position comes immediately after the decimal point, and has the value **7**, which means $^7/_{10}$ths of a dollar. The pennies position follows on the right, and has the value **5**, which means $^5/_{100}$ths of a dollar. The values in those two positions sum to **75** cents, or **$0.75**. The whole number part **$238** is to the left of the decimal point.

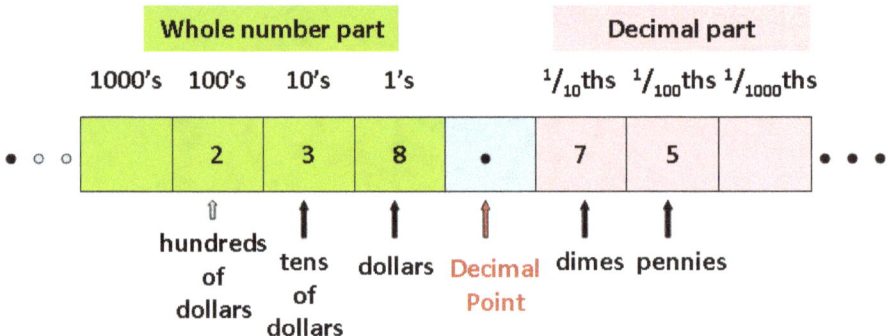

The positions of the digits that follow the decimal point are called **decimal places**. In everyday use, we write dollar amounts to **2** decimal places, as shown in the graphic. In the graphic, **7** is in the **first** decimal place, and **5** is in the **second** decimal place. However, in mathematics a decimal number must be expressed in as many decimal places as necessary to accurately represent its value.

Let us look at other examples of decimal numbers.

We write the number $8^1/_2$ in decimal as **8.5**. Again, the dot following **8** is the decimal point. The **5** after the decimal point represents $^5/_{10}$, which is equals $^1/_2$. The number **8.5** has one decimal place.

We write $8^1/_4$ in decimal form as **8.25**. The **2** in the first decimal place represents $^2/_{10}$, and the **5** in the second decimal place represents $^5/_{100}$. When we add $^2/_{10}$ and $^5/_{100}$, we get $^{25}/_{100}$, which is $^1/_4$. The number **8.25** has two decimal places.

We write $120^1/_8$ in decimal form as **120.125**. The **1** in the first decimal place represents $^1/_{10}$, the **2** in the second decimal place represents $^2/_{100}$, and the **5** in the third decimal place represents $^5/_{1000}$. When we add $^1/_{10}$, $^2/_{100}$ and $^5/_{1000}$, we get $^{125}/_{1000}$, which equals $^1/_8$. The number **120.125** has three decimal places.

9.3 Converting a Fraction to Decimal

In our everyday interactions, we most often present numbers in decimal form. Money is always in decimal form, as we have already seen. For example, we usually write **three dollars and fifty cents** as **$3.50**, not $3^1/_2$. We typically report the measures of the performance of athletes in decimal form. When we talk about a baseball player batting three hundred, what we really mean is that his batting average is 0.**300**. Quantity and size are also often presented in decimal form. Look at the label on a bottle of water and you may, for example, see the volume of its content presented as **16.9** fluid oz. Thus, it is important to know how to convert a fraction to decimal form.

Converting a fraction to decimal form is quite simple. Let us look at an example using the following division problem: **30 ÷ 8**. We know that the result, in mixed number form, is $3^3/_4$. To get the result in decimal form we use **long division**. The following example shows the steps involved.

Example 1: **30 ÷ 8**

Converting **30 ÷ 8** to decimal form using the long division method:

```
          3.75
    8 ) 30
        -24
          60
        - 56
          40
        - 40
           0
```

- We divide 30 by 8 and get the whole number quotient (3).
- We multiply 3 by 8 and get 24.
- We subtract 24 from 30 and get 6. Since 6 is less than the divisor 8, we are through with the whole number part of the quotient.
- We place a decimal point after 3. (We have **3.**)
- We multiply the remainder 6 by 10 and get 60.
- We divide 60 by 8 and get 7. We write 7 after the decimal point. We have **3.7**; the first decimal place is occupied by (7).
- We multiply 7 by 8 and get 56.

- We subtract 56 from 60 and get 4.
- We multiply 4 by 10 and get 40.
- We divide 40 by 8 and get 5.
- We write 5 after 7, and 5 occupies the second decimal place.
- We multiply 5 by 8 and get 40.
- Subtracting 40 from 40 gives zero, so we have completed the division.

The result in decimal form is: **3.75**.

Note that when we reach a remainder of the integer division, we write a decimal point, and then keep multiplying each successive remainder by 10 and dividing until the remainder becomes zero, or until we reach a specified number of decimal places. If we multiply a remainder by 10 and the resulting number is still less than the divisor, we place a zero in the corresponding decimal place and multiply by 10 again, then continue the division.

Note: Instead of saying explicitly: "multiply the remainder by 10", as we did in the example above, we sometimes say: "append, or add, zero to the remainder". In this case, the two phrases mean the same thing, they are just different uses of language. For example, multiplying **6** by **10** gives **60**, and appending zero to **6** also gives **60**.

Why do we multiply the remainder by **10** before dividing to get the value of the next decimal place? Let us look at the previous example: **30 ÷ 8**. The remainder is **6**. Multiplying **6** by **10** to get **60** allows us to continue the division process to get a value for the first decimal place. Since the first decimal place is $^1/_{10}$th of the **1**'s position, we are essentially performing:

$$6 \times 10 \times \, ^1/_{10} = 6.$$

The process gets us to the correct value for each decimal place, and the dividend we started with remains the same.

Let us look at another example.

Example 2: 10 ÷ 3

Converting **10 ÷ 3** to decimal form using the long division method:

```
        3.333
    3 ) 10
       - 9
        10
       - 9
        10
       - 9
        10
       - 9
        10
```

- 10 ÷ 3 = 3 r 1
- Write (3.) The dot is the decimal point.

- Multiply the remainder (1) by 10 and get 10, and continue the division.
- 10 ÷ 3 = 3 r 1
- Write (3.3) We have accounted for the first decimal place.
- Multiply the remainder (1) by 10 and get 10, and continue the division.
- 10 ÷ 3 = 3 r 1
- We write (3.33) We have accounted for the second decimal place.

You can see that the process repeats without end.

In this example, since the division process repeats without end, if we continued on we would get 3.3333333... Interminable division is a common occurrence in conversion to decimal form. (Try converting 10 ÷ 6 to decimal form.) Dividing indefinitely is unrealistic, so we stop after a specified number of decimal places.

It is important to understand that the decimal form is simply another way to write a number. Like other numbers, decimals can be added, subtracted, multiplied and divided.

A whole number can be written in decimal form by simply placing a decimal point after the number, followed by zeros in the decimal places. For example, **8.0** is the decimal representation, to one decimal place, of the whole number **8**. When we work in the realm of decimals, we write every number in decimal form. For integers we place zeros in the decimal places.

9.4 Adding Decimals

We add decimals the same way we add integers, right to left. We do **carry over** the same way we do in integer addition.

We must line up the decimal places, so that a decimal place in one number gets added to the **same** decimal place in another number.

> *Example 1*: Add **8.75** and **4.9**.
>
> We write:

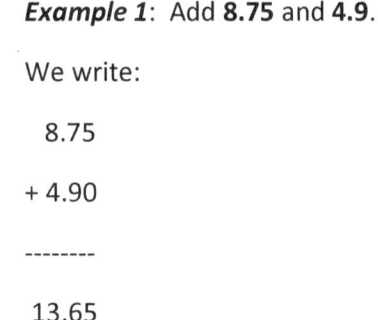

Note that **4.9** is written as **4.90**, so that it too has two decimal places, making it easy to line up decimal places.

- 0 + 5 = 5: therefore, 5 goes to the second decimal place.
- 7 + 9 = 16: therefore, 6 goes to the first decimal place. We carry over 1.)
- 8 + 4 = 12, plus the carry over 1 = 13. This is the integer part.

Final result: 13.65. Note that the final result also has two decimal places.

Example 2: Add **8.953**, **200.62**, and **93.1**

8.993

200.620

+ 93.100

302.713

- 3 + 0 + 0 = 3: therefore, 3 goes to the third decimal place.
- 9 + 2 + 0 = 11: therefore, 1 goes to the second decimal place. We carry over 1.
- 9 + 6 + 1 = 16, plus the carry over 1 = 17: therefore, 7 goes to the first decimal place. We carry over 1. We have completed the decimal places.
- We write the decimal point.
- We do the addition of the whole number part as usual, taking into account the carryover of **1** from the first decimal place addition.

9.5 Subtracting Decimals

We subtract decimals the same way we subtract integers, right to left. We **borrow** the same way we do in integer subtraction. Again, we must make sure that decimal places are lined up, so that a decimal place in one number is subtracted from to the **same** decimal place in another number.

Let us look at an example.

Example 1: Subtract 23.8972 from 38.221.

We write:

38.2210

− 23.8972

14.3238

- We borrow 1 from the third decimal to complete the fourth decimal place subtraction. The fourth decimal place gets 8.

- We again borrow 1 from the second decimal place to complete the third decimal place subtraction. The third decimal place gets 3.
- We again borrow 1 from the first decimal place to complete the second decimal place subtraction. The second decimal place gets 2.
- We again borrow 1 from the units position to complete the first decimal place subtraction. The first decimal place gets 3.
- We now complete the subtraction of the whole number part as usual, taking into account, the previous borrow of **1** from the units position. The result is **14.3238**.

9.6 Multiplying Decimals

We multiply decimals the same way we multiply integers. However, before we perform the multiplication, we count the total number of decimal places in both numbers being multiplied. The number of decimal places in the product is equal to the sum of decimal places in the two numbers.

> ***Example 1***: Multiply **12.1230** by **11.56**

In this example we are multiplying one number with 4 decimal places by another with 2 decimal places. Their product will have 6 decimal places. Since we know that the result has 6 decimal places, we can ignore the decimal points and treat both numbers as whole numbers, multiplying them as: **121230 × 1156**. After we have the result of the multiplication, we count six digits from **right to left** and place a decimal point to the left of the sixth digit.

Let us work through this example:

```
        121230
      ×   1156
      --------------
        727380
       606150
      121230
     121230
      --------------------
     140141880
```

The final decimal result is **140.141880**.

9.6.1 Multiplication of a Decimal by 10, 100, etc.

Remember that the first decimal place is a multiple of $^1/_{10}$. When we multiply a decimal by 10, we promote the number in the **first decimal place** to the **1**s position. This is the same as moving the decimal point one place to the right.

Since the second decimal place is a multiple of $^1/_{100}$, multiplying a decimal by 100 (that is, by 10 a **second** time) promotes the number in the **first decimal place** to the **10**s position and the number in the **second decimal place** to the **1**s position. This is the same as moving the decimal place two positions to the right. And so on.

9.7 Dividing Decimals

When we divide two decimal numbers, we first convert the divisor to a whole number. Then we perform the division following the same technique we previously used to get a decimal quotient.

Here is how we do the conversion. We count the decimal places in the divisor and move the decimal point that count of places to the right. The divisor is now a whole number. Next, we move the decimal point in the dividend the same count of places to the right, or we append a zero for each decimal place that is not present. The process amounts to repeatedly multiplying **both** numbers by 10 until the **divisor** has no more decimal places.

Let us look at an example.

> ***Example 1***: Divide 1080.6 by 0.12

- The divisor has 2 decimal places. We move the decimal point in the divisor 2 places to the right. The divisor becomes 12.
- Since the dividend has only one decimal place, we move its decimal point one place to the right and append a zero to account for the second decimal place. The dividend becomes 108060.
- Now we perform the division using the long division technique.

The result is 9005. (Confirm this result by working the division out yourself.) Remember that we must move the decimal point the same number of places to the right in both the divisor and the dividend.

9.8 Precision of Decimal Places

The more decimal places a decimal has the more precise the number is. The two decimals: **1.988** and **1.98** look similar, but **1.988** is closer to **2** than **1.98**. Thus **1.988** is bigger than **1.98**. We see this clearly when we multiply both numbers by 1000. It moves the decimal point 3 places to the right.

$$1.988 \times 1000 = 1988$$

$$1.98 \times 1000 = 1980$$

Needless to say, 1988 > 1980. We have a difference of 8. If we multiply both numbers by 10,000, we get 19880 and 19800. The difference has grown to 80! Thus, if we want to be as precise as possible when working with decimals, we should retain as many decimal places as practical.

9.9 Review Problems

Solve the following problems. Do not use a calculator.

i. $23.8355 + 192.866 + 10 =$ _____

ii. $14.862 \div 1.2 =$ _____

iii. $17.893 \times 23.0 =$ _____

iv. $1009.237 - 699.883 =$ _____

v. $0.998 - 0.99 =$ _____

vi.   A man's stride length is 2.8 feet. How many strides does he take to cover 4928 feet?
      Answer: _____

vii. A shopper paid \$10.21 for 1.79 pounds of coffee. What is the coffee's price per pound?
      Answer: _____

Chapter 10 - Rounding

10.1 Our Focus

In this chapter, we will discuss the rules for reducing the number of decimal places in a decimal number. We will also discuss how to round integers.

10.2 Rounding a Decimal

Sometimes we need to reduce the number of decimal places in a decimal number. For example, we write dollar amounts in two decimal places. However, when we divide $10 by 8, the result is $1.125. What do we do with the third decimal place?

There are three approaches for reducing decimal places.

One approach is **truncation**. In truncation, we simply cut off the decimal places we don't want. Truncating $1.125 to two decimal places, results in $1.12.

Another approach is **rounding**. We round **up** or round **down**.

Let us round $1.125 to **two** decimal places. We look at the **third** decimal place, which is the cut off. If the digit at the cutoff is 5 or larger, we add 1 to the digit to its left. In $1.125, the third decimal place has 5, so we add 1 to the number in the second decimal place: 2, and get $1.13. We round **up**.

Similarly, when we round 8.8991499 to two decimal places, we round **up**. We get 8.90, because we add 1 to 9 in the second decimal place and get 10 and carry over 1 as required in addition.

Rounding 8.8991499 to four decimal places gives 8.8991. The number in the fifth decimal place is 4, which is less than 5, so we don't add 1 to the digit in the fourth decimal place. In this case, we round **down**.

When we round a decimal to the nearest whole number, we consider the digit in the first decimal place. If it is 5 or larger, we add 1 to the units position. Rounding 100.5 to the nearest whole number gives 101.

10.3 Rounding Integers

Rounding applies to more than just decimals. Sometimes we round integers as well. We can round to the nearest 10, to the nearest 100, and so on.

When we round 55 to the nearest 10, we get 60. The **nearest ten** refers the **adjacent numbers which are both multiples of 10.** For 55, the nearest 10 is either 50 or 60. Since 55 is halfway to 60, we round up to 60. On the other hand, when we round 54 to the nearest 10, we get 50. That is, we round down because 54 is closer to 50 than it is to 60.

Similarly, the **nearest 100** means **adjacent numbers which are both multiples of 100**. When we round 345 to the nearest 100, we get 300. On the other hand, when we round 355 to the nearest 100, we get

400. When we round 35 to the nearest 100, we get zero. We round down because 35 is closer to zero that it is to 100.

10.4 Review Problems

Solve the following problems. Round decimal results of the first two problems to 3 decimal places.

    i. $213.309 - 93.6412 =$ _____

    ii. $93.425 + 63.5588 =$ _____

    iii. $64.80 \div 0.08 =$ _____

    iv. $12.34 \times 1.18 =$ _____

    v.   Round 67.167 to two decimal places. Answer: _____

    vi.   Round 876. 149 to one decimal place. Answer: _____

    vii.   Round 856 to the nearest 100. Answer _____

Chapter 11 – Ratios and Percentages

11.1 Our Focus

Ratios and **Percentages** are closely related concepts. They both provide an insightful way to compare two values. In fact, percentages are ratios expressed on a scale of one hundred (100). In the first part of this discussion, we will cover ratios. Then we will apply what we learn to percentages.

11.2 Ratios

Imagine that two friends, Jim and Brian, spend 12 months in California. During the 12-month stay, Jim spends $14,870 and Brian spends $16,380. Looking at the expenses, we can tell that Brian's year was more expensive than Jim's by $1510. The difference gives us meaningful information, but the ratio of the expenses gives us an even more telling insight.

Let us take the ratio of Jim's expenses to Brian's. It means we use Brian's expenses as the basis for comparison. To get the ratio, we divide $14,870 (Jim's expense) by $16,380 (Brian's expense). The result of the division is 0.91 (rounded to 2 decimal places.) The result tells us that during the 12-month period, Jim spent 91 cents ($0.91) for every dollar Brian spent.

Let us do the reverse by taking the ratio of Brian's expenses to Jim's. Now, Jim's number is the basis for comparison. We divide $16,380 by $14,870. The result of the division is 1.10 (rounded to 2 decimal places.) The result tells us that Brian spent $1.10 for every dollar Jim spent during the 12-month period.

Ratios allows us to make meaningful comparisons between **similar** values. In the example we just discussed, the values were both 12-month expense totals.

We are going to stay with Jim and Brian but modify the circumstances and expenses a little. Jim stays in California for 12 months and spends $14,880. However, Brian stays for only 7 months and spends $9,030.

The values are no longer similar, because they are not based on the same duration of stay. One way to make the values similar is by averaging the expenses over the months of stay. That gives us average monthly expenses for both men. We now have a similar basis for comparison.

The average monthly expense for Jim is: $14,880 divided by 12, which equals: $1240. Brian's average monthly expense is: $9,030 divided by 7, which equals: $1290. The ratio of Jim's monthly expenses to Brian's is:

    1240 ÷ 1290 = 0.96 (to 2 decimal places.)

The ratio tells us that as a monthly average, Jim spent 96 cents ($0.96) for every dollar Brian spent.

Again, we can take the reverse ratio: Brian's to Jim's. We get:

    1290 ÷ 1240 = 1.04 (to 2 decimal places.)

This ratio tells us that as a monthly average, Brian spent $1.04 (one dollar and 4 cents) for every dollar Jim spent.

Let us summarize the key points of the discussion:

- We use a ratio to compare two similar values.
- We get the ratio by dividing one value by the other.
- The divisor serves as the basis for comparison.

11.3 Percentages

We learned from the previous examples that we calculate a ratio as a fraction, expressed in decimal form. In the last example, one ratio was 0.96, the other was 1.04. In everyday use, we often report a ratio as a **percentage** by multiplying it by 100. This is where the 100 scale we referred to earlier comes from. In percentage terms, 0.96 is 96 **percent** (96%) and 1.04 is 104 **percent** (104%).

The word **percent** means **per hundred**. "Cent" in per**cent** is a contraction of the Latin word "centum", which means hundred. The symbol for percent is: **%**.

Again, to report a ratio as percent, we multiply the decimal result of the division by 100 and add the % sign, as we did above. We must round the decimal result to the appropriate number of decimal places. Consider the reverse of this. If a ratio is reported as percent, dividing it by 100 gives us the decimal value of the ratio.

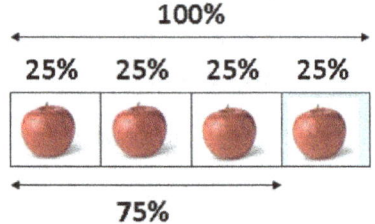

Let us take a closer look at percentages using the graphic above. The graphic shows a total inventory of **4** apples. If we sell all four apples, we have sold 100% of the inventory. How so? Well, the ratio of the number of apples we sold to the number of apples in the original inventory is $^4/_4$ = **1**. Multiplying **1** by **100** and applying the percent sign gives 100%. It is that simple!

In the graphic, each apple represents a $^1/_4$, or **0.25**, of the inventory. If we multiply **0.25** by **100**, we get **25**%. So, we say that each apple represents **25**% of the inventory of **4** apples. Thus, another way of saying we sold **1** out of **4** apples is: we sold **25**% of our inventory of **4** apples.

If we sell one apple, we have three apples left. The ratio of **apples left** to the original inventory is $^3/_4$, or **0.75**. Multiplying 0.75 by 100 gives 75%.

If we consider the original inventory of four apples as 100% and subtract 25% (1 apple sold) from 100% (original inventory of 4), we arrive at 75%. If we now add 1 apple (25%) to restore the inventory to 4 apples, we are back to 100% of the original inventory. We can see that percentages can be added and subtracted as long as they are calculated using the same basis.

If we had started with an inventory of 3 apples, the 100% basis would be 3 apples. If we sold 1 apple, the percentage sold would be 1 out of 3, that is: $^1/_3$ = 0.33 (to two decimal places.) In percentage terms, it would be 33%. The two apples left would be 67% of the original inventory. The sum of 33% and 67% is 100%.

11.4 Percentages in Everyday Situations

Let us look at how we apply percentages in everyday situations. We will consider a number of different examples.

Example 1: Jim pays $45 for a pair of shoes on sale. The shoe goes off sale and his buddy, Pablo, pays $50 for a pair. The ratio of the amount Jim paid to the amount Pablo paid is $^{45}/_{50}$. On a percentage scale, Jim paid 90% of what Pablo paid. We get this as follows:

$^{45}/_{50}$ = 0.90, and 0.90 × 100% = 90%.

On the other hand, the ratio of what Pablo paid to what Jim paid (the reverse of the previous ratio) is: $^{50}/_{45}$ = 1.11 (to two decimal places.) In percentage terms: 1.11 × 100% = 111%, telling us that Pablo paid 111% of what Jim paid.

In the second ratio we used Jim's payment as the basis for comparison. Since Jim paid **less** than Pablo, the resulting ratio is greater than 1, and the resulting percentage is **greater** than 100%.

Example 2: If we leave a tip of $18 for a restaurant bill of $98, the ratio of the tip to the bill is the fraction $^{18}/_{98}$ = 0.184 (to three decimal places.) To convert to percent, we multiply the result by 100:

0.184 × 100 = 18.4%.

Example 3: If a company discounts a product that normally sells for $56 by 17%, the new sales price is 83% of the original. The new sales price and the discount amount sum up to the original sales price, because 17% and 83% sum up to 100%.

Original price = $56 (100%)

83%	17%
Discount Price = $M	Price Decrease

$M = $^{83}/_{100}$ × $56 = $46.48

The new sales price is $^{83}/_{100}$ of $56, given by:

$^{83}/_{100}$ × $56 = $46.48

(Remember that "**of**" means "multiply by".) The new price: $46.48 is 83% of the original price: $56. The difference between $56 and $46.48 is $9.52. It is the 17% the store gave away as discount.

Example 4: A store experiences an unusually high demand for an item and increases the item's price by 17%. The items original price was $46.48. If we use $46.48 as the basis for comparison, then it represents 100%. The new price will be 117% of the original price. The graphic below illustrates this.

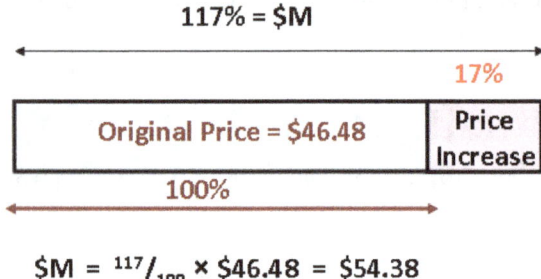

$$\$M = {}^{117}/_{100} \times \$46.48 = \$54.38$$

We calculate the new price as follows:

$${}^{117}/_{100} \times \$46.48 = \$54.38; \text{ rounded to 2 decimal places.}$$

The new price: $54.38 is **117% of the original price**: $46.48.

We can take the reverse ratio of the original price ($46.48) to the new price ($54.38) and see what it tells us. It is:

$${}^{46.48}/_{54.38} \times 100 = 85.5\% \text{ (to one decimal place.)}$$

It tells us that the original price is 85.5% of the new price.

Example 5: An item is sold at 15% discount. The discount price is $65. What is the original price? The graphic below illustrates this problem.

Original price = $M (100%)

	15%
Discount Price = $65	Price Decrease

85%

$$\$M = {}^{100}/_{85} \times \$65 = {}^{65}/_{0.85}$$

The first thing we must understand is that the 15% discount came off the original price, not the discount price of $65. Therefore, we cannot add 15% of $65 to get the original price.

The discount price, $65, is 85% of the original, as the graphic illustrates. In the graphic, the original price is shown as an unknown value $M. We know that 85% of $M equals $65. Now we determine the value of $M.

85% = ${}^{85}/_{100}$ = **0.85**, therefore **0.85** × $M = $65. From this,

M = $65 ÷ **0.85** = ${}^{\$65}/_{0.85}$ = 76.47.

$M = $76.47 (to two decimal places.)

In this example, $65 is 85% of the price we are looking for, meaning that the new price is greater than $65. To get it we divide $65 by 0.85.

Problems involving percentages come in many flavors. Like any problem that we solve, we must understand the facts. Particularly, we must understand the basis for the percentages involved. Once we are clear about those facts, we apply the appropriate elements of the concepts we have discussed to determine the desired result.

11.5 Review Problems

i. Mary saved 10% of her paycheck each month for 12 months. At the end of the period she had accumulated $2880 in her savings account. What was Mary's monthly pay?

ii. Cecil and his friends had lunch at a local restaurant. Their bill was $80.00. They left $15 as tip. What percentage of their food bill was the tip?

iii. A shoe store found itself with too much inventory in one model. The store discounted the shoe by 18% and sold 63 pairs. If the revenue from the sale was $3024, how much was the original price of the shoe?

iv. A mother's height is 85% of her son's. Her son is 6 feet 8 inches. How tall is the mother?

v. Marie has $12.75 to spend at the movies. Her ticket costs $9.00. What percentage of her money is left for snacks after she pays for the ticket? Round to the nearest whole number.

vi. A crowd of 640 students attended the school's basketball game. The number represents 40% of the student body. How many students are in the school?

Chapter 12 – Variables Revisited

12.1 Our Focus

We previously discussed variables, but at a high level. That basic understanding allowed us to consider variables in our discussion of fractions. In this chapter, we will delve more deeply into variables. We already know that we use variables in operations in the same way we use numerals. Here we will discuss more complex expressions involving variables. We will look at examples where we mix variables and numerals together in the same expression.

12.2 Simplifying a Variable Expression - A Closer Look

Variable expressions typically involve multiple variables, products of variables, and numerals. The purpose of simplifying such an expression is to consolidate **common terms** and make the expression simpler, clearer, and easier to evaluate. Common terms are also referred to as **like** terms. We consolidate by performing the arithmetic operations in the expression that apply to each set of common terms.

What are common terms?

- Numerals are common terms.
- Terms that contain numeric multiples of the same variable are common terms.
- Terms that contain numeric multiples of the same product of variables are common terms.
- Terms that contain the same exponent of the same variable or its numeric multiples are common terms. (We will discuss exponents later. Exponents are not shown in the graphic.)

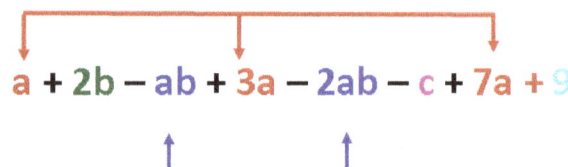

These Common Terms get consolidated

$$a + 2b - ab + 3a - 2ab - c + 7a + 9$$

These Common Terms get consolidated

2b, c, and 9 are by themselves

Let us consolidate the expression in the graphic as our first example.

> ***Example 1***: $a + 2b - ab + 3a - 2ab - c + 7a + 9$

Going from left to right, we consolidate the a terms: $a + 3a + 7a = 11a$

Next we consolidate the ab terms: $- ab - 2ab = - 3ab$

The remaining terms can't be consolidated with any other terms. Our result is:

11a + 2b − 3ab − c + 9

Let us work on another example.

Example 2: $2a - 3(b - a + 2)$

In this example the variable **a** appears in two different terms: in the product **2a** and inside the grouped part. We can consolidate the two appearances of **a**. The variable **b** appears only once, inside the grouped part. No additional consolidation of **b** can be done.

Before we can consolidate we must remove the grouping. To do so, we first multiply **3** through the grouped part. Then, <u>because the entire grouped part is being **subtracted** from the operand **2a**</u>, we reverse the addition and subtraction operations **inside** the grouped part. That is, inside the grouped part addition changes to subtraction and subtraction changes to addition. Now we remove the parentheses.

Let us do this in steps:

- We multiply the grouped part through with 3:

    2a − 3(b − a + 2) becomes: **2a − (3b − 3a + 6)**

- **Inside** the grouped part, we change subtraction to addition, and addition to subtraction and remove the parentheses:

    2a − **(3b − 3a + 6)** becomes: **2a − 3b + 3a − 6**;

- We add **2a** to **3a** and get **5a**.

- The resulting simplified expression is: **5a − 3b − 6**.

Note: *We reverse addition and subtraction operations inside the grouped part **only when** the whole grouped part is the right operand of a subtraction. We will see the explanation for this later in the discussion.*

Let us simplify another expression.

Example 3: $2a + 3bc - 2(a - b) + 7(bc - 2) + 13b + 8$

The following are the like terms in this expression:

- The terms that have the variable a.
- The terms that have the variable b.
- The terms that have the bc variable product.
- The numerals.

Again, we simplify the expression by consolidating multiple appearances of like terms. As we did in the previous example, we multiply **2** and **7** through their related grouped parts. Because the group **2(a − b)** is a **subtracting** operand, we change the subtraction inside the group to addition before removing the parentheses. Let us simplify the expression step-by-step:

- We multiply through the grouped parts:

2(a − b) becomes: (2a − 2b), and

7(bc − 2) becomes: (7bc − 14)

- We insert these back into the expression and it becomes:

$2a + 3bc − (2a − 2b) + (7bc − 14) + 13b + 8$

- We reverse the operation inside the grouped part (**2a − 2b**) to (**2a + 2b**), because it is a subtracting operand, and remove the grouping symbols. *Since the grouped part (7bc − 14) is not a subtracting operand, we do not change the subtraction inside it to addition.* We get the following intermediate expression:

$2a + 3bc − 2a + 2b + 7bc − 14 + 13b + 8$

- We perform the consolidation:

    i.   **2a − 2a** is zero. Those terms cancel each other out and are eliminated.

    ii.   **3bc** + **7bc** equals **10bc**.

    iii. **2b** + **13b** = **15b**.

    iv.   (**−14**) + **8** = **8 − 14** = **-6**.

Our final result is: **10bc + 15b − 6**.

Let us simplify one more expression.

Example 4: $12a − 3b − (2a + 4c) ÷ 2 − (12b + 4c) ÷ 3$

In this expression, the two grouped terms that have been highlighted are being divided. Just as we can multiply through a grouped term with its multiplier, we can also divide through a grouped term with its divisor. Let us simplify the expression step by step. First we perform division.

$(2a + 4c) ÷ 2$

$= (2a ÷ 2 + 4c ÷ 2)$

$= (a + 2c)$; [this is equivalent to $\frac{1}{2}(2a + 4c)$]

And:

$(12b + 4c) ÷ 3$

$= (12b ÷ 3 + 4c ÷ 3)$

$= (4b + \frac{4c}{3})$; [this is equivalent to $\frac{1}{3}(12b + 4c)$]

We replace the two grouped terms with the intermediate results.

$12a − 3b − (2a + 4c) ÷ 2 − (12b + 4c) ÷ 3$ becomes:

$$12a - 3b - (a + 2c) - (4b + {}^{4c}/_3)$$

$= 12a - 3b - a - 2c - 4b - {}^{4c}/_3$; (note that addition in the grouped parts have reversed to subtraction.)

Now, we consolidate common terms:

- $12a - a = 11a$;
- $-3b - 4b = -7b$
- $-2c - {}^{4c}/_3 = -{}^{6c}/_3 - {}^{4c}/_3 = -{}^{10c}/_3$

The final result is: $11a - 7b - {}^{10c}/_3$.

12.2.1 Why We Reverse Signs Inside Subtracting Group

Why do we reverse the operations inside subtracting grouped parts when we remove the parentheses? Remember that subtraction is really the addition of a negative number. With this in mind, let us revisit the expression we used in *Example 2*.

Original expression is:

$2a - 3(b - a + 2)$

We said earlier that subtraction is the addition of a negative operand. Consequently, we can rewrite the expression, changing subtraction of the grouped part to addition of a negative operand:

$2a + -3(b - a + 2)$.

Now, we multiply through the grouped part with the multiplier (**-3**). Following our rules for multiplying positive and negative numbers, we get:

$2a + (-3b + 3a - 6)$

Note that the subtraction and addition in the grouped part of the original expression have reversed as a result of multiplying by a negative.

When we remove the parentheses, we get:

$2a + -3b + 3a - 6$

$= 2a - 3b + 3a - 6$

And, consolidating like terms, we get the same result as before:

$5a - 3b - 6$.

Reversing addition and subtraction operations inside a **subtracting** grouped part is a short cut. It saves us the steps of multiplying through with a negative multiplication factor.

In the expression: $2a - (b - a + 2)$, the grouped part has a multiplier of **1**. We simplify $2a - (b - a + 2)$ by simply reversing addition and subtraction inside the grouped part and removing the grouping symbols, since multiplying by **1** does not change anything. We get:

$$2a - b + a - 2 = 3a - b - 2$$

We can re-write the original expression as follows:

$2a + -(b - a + 2)$; that is, change from subtraction to the addition of a negative operand. [The more explicit form is: $2a + -1(b - a + 2)$.]

In this form, the grouped part has the multiplier (-1). Multiplying by (-1) does not change bigness, but it reverses sign. Thus, when we multiply through the grouped part by (-1) we get:

$$2a + -b + a - 2$$

$= 2a - b + a - 2 = 3a - b - 2$, which brings us to the original result.

12.2.2 Review Problems

Simplify the following expressions:

i. $2b - 2a - 17 + 4b + 4a + 2(a - 28) =$

ii. $12a + (-3b) - 2a + 4c + 3(7b + 4c) =$

iii. $12a + 3b - 2(a - 32) - 3(7b + 4a) - 1 =$

iv. $12a - (-3b) - 2a + 4c - 3(7b + {}^4/_3 c) =$

v. $12a - 3b - (12a + 4c) \div 6 + (15b + 2c) \div 3 =$

vi. $32 + (2a + 4c - 12) \div 2 - (12a + 16c) \div 4 =$

12.3 Evaluating a Variable Expression

Let us start with the following simple expression:

Example 1:   a + b.

If we are told that **a** = 10 and **b** = 5, we replace **a** and **b** in the expression with their assigned numeric values.

a + **b** becomes: 10 + 5.

Because we now have numeric values for **a** and **b**, we are able to **evaluate** the expression and get a numeric result: **15**.

Let us evaluate the next expression using the numeric values:

a = 10, and b = 5.

Example 2:   25 + 3a − 2(a + b) − 4b

Substituting 10 for **a** and 5 for **b**, we get:

$$25 + 3 \times 10 - 2(10 + 5) - 4 \times 5.$$

Remember that 2(10 + 5) means 2 × (10 + 5).

Following our rules for evaluating an expression, we get

$$25 + 30 - 2 \times 15 - 20$$

$$= 25 + 30 - 30 - 20 = 5$$

Using the same values: **a** = **10** and **b** = 5, let us look at another expression, this time involving division:

Example 3: (4a − 2b) ÷ 2b, which we re-write as:

$$\frac{4a - 2b}{2b}$$

$$= \frac{4 \times 10 - 2 \times 5}{2 \times 5}$$

$$= \frac{40 - 10}{10}$$

$$= \frac{30}{10} = 3$$

Let us evaluate one more expression, given **a** = 2, **b** = 11:

Example 4: |2a − 3b − 2(b − 2a)|

In this example, we want the absolute value of the result from the evaluation of the expression. Substituting the values assigned to **a** and **b** into the expression, we get:

$$|2 \times 2 - 3 \times 11 - 2(11 - 2 \times 2)|$$

$$= |4 - 33 - 2(11 - 4)|$$

$$= |4 - 33 - 14| = |-43| = 43$$

In each example, we first replaced the variables with their assigned numeric values. We then evaluated the resulting numeric expression by following the order of operation rule.

Of course, we could have simplified the expressions before performing the evaluation. To this end, let us go back to the expression in ***Example 2***:

$$25 + 3a - 2(a + b) - 4b$$

Simplifying the expression gives:

$$25 + a - 6b$$

Applying the assigned values: **a** = **10**, **b** = **5**, we get:

$$25 + 10 - 30 = 5$$

Clearly, the evaluation of the simplified expression is much more straight forward than the evaluation of the original expression.

If we get an expression that can be simplified, we are better of simplifying it first before evaluating. Doing so can make the evaluating easier, especially if the variables are assigned fraction values.

12.3.1 Review Problems

i.   Evaluate the following variable expressions, given a = 2, b = 4, c = 3:

1. $(a + b + c) \div (c - a) =$

2. $b(c - 5a) + 10c \div a - 5b =$

3. $c(a + b) - 2(b - a - c) + 36 =$

ii.   Evaluate the following variable expressions, given a = $^1/_2$, b = $^1/_4$, c = $^1/_3$:

1. $3(2a - b + c) - 3(a - b + c) =$

2. $12(b + 2c) \div 8a =$

3. $|-a \div 2c| =$

Chapter 13 - Single Variable Equations

13.1 Our Focus

In the introduction to variables, we said that the value of a variable is determined based on the facts of the problem we are presented with. In this discussion we will learn how to use available information to determine the values of variables in simple equations involving one variable.

13.2 The Form of an Equation

Let us start with the following example:

> ***Example 1***: $5 - x = 2$. (We have named the variable **x**.)

Example 1 is an algebraic statement called an **equation.** It is a simple equation involving one variable: **x**.

An equation has an **equal sign** with an expression on either side of it. In this example, the expression on the right is quite simple. It is just the numeral (**2**). The equality tells us that there is a value of the variable **x** such that $5 - x$ equals **2**. We solve the equation by finding that value of **x**.

The graphic below illustrates the equation. It is simple enough that we can solve it by inspection. Clearly, the solution is: **x = 3**.

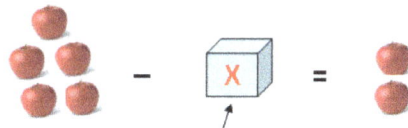

We must be removing 3 apples!

X = 3 apples

From our discussion of basic addition and subtraction, we know that if:

$a + b = c$, then: $a = c - b$, and: $b = c - a$.

For example, given: **a** = 7, **b** = 10,

> $a + b = c = 17$.

> $a = c - b = 17 - 10 = 7$.

> $b = c - a = 17 - 7 = 10$.

Let us apply this knowledge to solve the following examples.

> ***Example 2***: $2 + b = 5$.

> Therefore: $b = 5 - 2$.

> The solution is: **b** = 3.

> ***Example 3***: $b - 3 = 5$

Therefore: **b** = 5 + 3.

The solution is: **b** = 8.

Example 4: 4**b** − 9 = 11

Therefore: 4**b** = 11 + 9 = **20**.

We know from our discussion of basic multiplication and division that if:

ab = **c**, then: **a** = **c** ÷ **b**, and: **b** = **c** ÷ **a**.

(Remember that **ab** means **a** × **b**. Likewise **4b** means **4** × **b**.)

From: 4**b** = **20**, we get: **b** = 20 ÷ 4.

The solution is: **b** = **5**.

In order to solve for the variable in the examples we followed a process that allowed us to isolate the variable on one side of the equal sign. At that point, the value on the opposite side of the equal sign is the solution.

13.3 Structured Process for Isolating the Variable

We use the process of elimination technique as the more structured approach to solving a simple equation. Let us see how it works in the next example.

Example 1: 2 + **b** = 5

The following graphic illustrates the process of elimination technique.

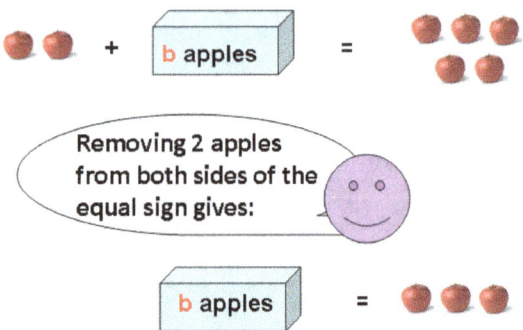

Following Smiley's suggestion and removing **2** apples from both sides of the equal sign, we determine right away that the box contains **3** apples.

We apply Smiley's idea to ***Example 1*** by subtracting **2** from both sides of the equation: 2 + **b** = 5. We get:

(left side) **2** + **b** − **2** = (right side) **5** − **2**.

The left side reduces to **b**; the right sides reduces to **3**; giving us the solution: **b** = **3**.

We can solve a simple equation as a mental exercise, or we can follow a structured process of elimination, like we just did. The objective is always to isolate the variable on one side of the equal sign.

The process of elimination may require adding, subtracting, multiplying and dividing values on both sides of the equal sign. When we perform an operation on one side of the equal sign, we must perform the same operation on the other side of the equal sign. In the example above, we subtracted **2** from **both** sides of the equal sign.

Let us use the process of elimination to solve another equation. This time we will formulate the equation from a word problem.

> **Example 4**: I have a number in my head. If you multiply the number by **4** and add **2**, you get **10**. What is the number?

First, because we don't know what the number is, we assign it a variable name: **h**. (Remember we can assign any alphabetic character.) The problem indicates that when you multiply **h** by **4** and add **2** you get **10**.

We write **multiply h by 4** as:

$$h \times 4 = 4h.$$

(Remember we write the numeral first.)

We write **Add 2 to 4h** as:

$$4h + 2.$$

We are told the sum equals **10**, so we write:

$4h + 2 = 10$; this is the resulting equation.

First, let us solve the equation as a mental exercise. What is the number that when added to **2** gives **10**? The answer is **8**, because **8** + **2** equals **10**. So, we know that **4h** equals **8**. Now we ask: What is the number that when multiplied by **4** gives **8**? The answer is **2**, because **2** × **4** = **8**. Therefore, **h** = **2**. We have solved it!

Next, let us use the step-by-step process of elimination technique to solve the equation.

We start with: **4h + 2 = 10**.

- We subtract **2** from left side of the equation and also from the right side:

  $4h + 2 - 2 = 10 - 2.$

  The result is: $4h = 8$.

- Now, we divide the left side by **4** and the right side also by **4**:

  $\frac{4h}{4} = \frac{8}{4}$, which is the same as:

Left side: $^{4 \times h}/_4 = {}^{1 \times h}/_1 = h$

Right side: $^8/_4 = 2$.

The left side reduces to **h** because of the common factor **4**. The right side reduces to **2**, giving us the solution: **h = 2**.

Again, the steps we followed allowed us to isolate **h** on one side of the equal sign while maintaining equality with the value on the other side. It does not matter on which side the variable is isolated.

Let us look at another example:

Example 5: 4h − 8 = h + 4

Again, we want to solve for **h** while maintaining equality between the left side and the right side of the equation. We notice that **h** appears on both sides of the equation, and there are numerals on both sides as well. To find the solution, our steps should lead to **h** by itself on one side of the equality.

- We add **8** on both sides:

4h − 8 + **8** = h + 4 + **8**

This eliminates **8** on the left side, since (**-8 + 8**) equals zero. Now, we have:

4h = h + 12

- We subtract **h** from both sides:

4h − **h** = h + 12 − **h**

On the right side, **h − h = 0**, leaving **12**. (Because any numeral, variable, or expression subtracted from itself gives zero.) On the left side, 4**h** − **h** = 3**h**.

The equation becomes:

3h = 12

- We divide each side by **3**. On the left, $^{3h}/_3$ reduces to **h**. On the right side $^{12}/_3$ gives **4**.

The solution is: **h = 4**.

Every time we reach a solution, we must go back to the original problem and substitute the solution in place of the variable. If the solution is correct, the equality will be maintained. Let us do this here. Our original expression is:

4h − 8 = h + 4

Replacing **h** with **4**, and evaluating both sides, we get:

4 × 4 − 8 = 4 + 4

16 − 8 = 8

8 = 8, which is true, telling us our solution is correct.

Let us solve one more equation:

Example 3: $\frac{3}{5}h + 15 = 2h + 5$

- The first step is to get rid of the fraction $\frac{3}{5}$. To do that we multiply **both sides** by **5**.

- On the left side we get:

$(5 \times \frac{3}{5} \times h) + (5 \times 15)$

$= 3h + 75$

- On the right side we get:

$5(2h + 5)$

$= 10h + 25$

We have changed the original equation to the following:

$3h + 75 = 10h + 25$

- We subtract **25** from both sides and get:

$3h + 75 - 25 = 10h + 25 - 25;$

which gives:

$3h + 50 = 10h$

- We subtract **3h** from both sides and get:

$3h + 50 - 3h = 10h - 3h$

The left side equals: **50** and the right side equals **7h**. So, we have:

$50 = 7h$

- We divide each side by **7**, and get: $\mathbf{h} = \frac{50}{7}$. (It doesn't matter if we write: $h = \frac{50}{7}$ or $\frac{50}{7} = h$.)

The solution is: $\mathbf{h} = \frac{50}{7}$, which, converted to a mixed number, is: $\mathbf{7\frac{1}{7}}$.

Now we test our solution. The original equation:

$\frac{3}{5}h + 15 = 2h + 5$

- We substitute $\frac{50}{7}$ for **h**. The left side becomes:

$\frac{3}{5} \times \frac{50}{7} + 15$

$= \frac{30}{7} + 15$

$= 4\frac{2}{7} + 15 = 19\frac{2}{7}$

- The right side becomes:

$$2 \times {}^{50}/_7 + 5$$

$$= {}^{100}/_7 + 5$$

$$= 14^2/_7 + 5 = 19^2/_7$$

We have: $19^2/_7 = 19^2/_7$, which maintains the equality, telling us our solution is correct.

13.4 Review Problems

Find the answers for the following problems:

i. I have a number in my head. If you multiply the number by 3 and add 14 you get 41. Write the equation for the problem and solve for the number.

ii. Three boys went fishing. Together they caught 18 fish. One boy caught twice as many fish as the other. Write the linear equation for the problem and solve for the number of fish each boy caught.

iii. A young man is $^1/_2$ his father's age today. Ten years ago he was $^1/_3$ of his father's age. What is the young man 's age today?

iv. A hardware dealer purchased an inventory of 24 shovels at $14 each. He resold some of the shovels for $20 each and the remainder $18 each. He made a profit of $132. How many shovels did he sell for $18?

Chapter 14 - System of Linear Equations

14.1 Our Focus

In the previous chapter we discussed the steps for solving a simple equation with one variable or unknown. In this chapter, we will discuss how to solve systems of equations with two variables or unknowns. The equations we will discuss are called linear equations because they represent straight lines.

14.2 Solving a System of Linear Equations

A system of equations consists of two or more equations with the same variables or unknowns. The equations in a system have a common solution, if a solution exists. (Later, when we discuss the coordinate system, we will learn that the solution for a system of linear equations represents the coordinates of the point where the lines cross. But more on that later.)

As usual, we will start our discussion with an example.

   Example 1: I have two numbers in my head. Let us call the numbers **x** and **y**. When you multiply **x** by **2** and add **y**, the result is **8**. When you subtract **y** from **x**, the result is **4**. What are the numbers **x** and **y**?

First, we put the facts of the puzzle together to create equations. Since there are two variables, we need two equations. Let us consider the first set of facts: When you multiply **x** by **2** and add **y**, the result is **8**. From this statement we write:

   $2x + y = 8$

The second statement says: When you subtract **y** from **x**, the result is **4**. From the second statement we write:

   $x - y = 4$

Now we have two equations with two variables: **x** and **y**. The equations depend on each other because we need them both in order to find values for **x** and **y**. In other words, they form a system. The values should maintain equality in both equations. Because of this dependence, the equations are sometimes referred to as **simultaneous** equations; the solutions for **x** and **y** make both equations simultaneously true.

There are many strategies for approaching this problem. We will examine three of them. In each approach, we have the same objective: reduce one of the equations to a single variable equation and solve it using the procedure we learned previously. Once we know one of the values, we substitute it into either one of the two equations and solve for the value of the second variable.

The approach we will use in this example is the **addition method**. This method is based on the following simple truth: If you add the **left** sides of the two equations together, then add their **right** sides together, the resulting left side remains equal to the resulting right side.

Why the addition strategy for this case? We inspect the two equations and notice that on the left side of **Equation 1** the variable **y** is being added, and on the left side of **Equation 2** the variable **y** is being subtracted. That tells us that we can eliminate the variable **y** by adding the two left sides. This insight comes from the knowledge we gained from simplifying variable expressions.

Let us proceed. The equations are:

Equation 1: **2x + y = 8**

Equation 2: **x − y = 4**

Here are the steps for the addition method:

- Add the **left** side of **Equation 1** to the **left** side of **Equation 2**:

$(2x + y) + (x − y)$

$= 2x + y + x − y$; the **y**s eliminate each other.

$= 3x$

The result of adding the two left sides is: **3x**.

- Add the **right** side of **Equation 1** to the **right** side of **Equation 2**:

$8 + 4 = 12$

The result of adding the two right sides is: **12**.

Since the left side equals the right side, we have:

$3x = 12$; so **x = 4**.

Now we go to **Equation1** (we could have used **Equation 2**) and substitute **4** in place of **x**. We get:

$2 × 4 + y = 8$

$8 + y = 8$

$y = 0.$

Our solution is **x = 4, y = 0**.

Finally, we substitute these values into both equations and verify that equality is maintained:

Equation 1: $2x + y = 8$

$2 × 4 + 0 = 8$

$8 = 8$

Equation 2: $x − y = 4$

$4 − 0 = 4$

$$4 = 4$$

In the next example, we use the **subtraction** strategy. It is based on this fact: If we subtract the left side of *Equation 1* from the left side of *Equation 2* and then subtract the right side of *Equation 1* from the right side of *Equation 2*, the resulting left side remains equal to the resulting right side. When we use the subtraction method, we must make sure the left and right sides **of the same** equation are used in the subtractions. It is not important which equation we tag as **1** or **2**.

Example 2: Solve the following equations for **x** and **y**:

Equation 1: **4y + 2x = 16**

Equation 2: **3y + 2x = 3**

Why the subtraction method in this case? We notice that the term **2x** is being added in both equations. That tells us that if we use the subtraction method, the **2x** terms will cancel each other out.

Let us do it!

- Subtract the left side of *Equation 2* from the left side of *Equation 1*:

$$(4y + 2x) - (3y + 2x)$$

$$= 4y + 2x - 3y - 2x$$

$$= y$$

- Subtract the right side of *Equation 2* from the right side of *Equation 1*:

$$16 - 3 = 13$$

So, we have: **y = 13**.

As we did in the first example, we substitute **y = 13** into one of the original equations. It does not matter which one you choose. Let us use *Equation 2*:

$$3y + 2x = 3$$

$$3 \times 13 + 2x = 3$$

$$39 + 2x = 3$$

$$2x = -36$$

$$x = -18$$

Our solution is: y = 13, x = -18

Let us substitute these values into *Equations 1* & *2* to verify that the equality is maintained in both equations:

Equation 1: 4y + 2x = 16

$$4 \times 13 + 2 \times (-18) = 16$$

$$52 + (-36) = 16$$

$$52 - 36 = 16$$

$$16 = 16$$

Equation 2: $3y + 2x = 3$

$$3 \times 13 + 2 \times (-18) = 3$$

$$39 + (-36) = 3$$

$$39 - 36 = 3$$

$$3 = 3$$

We have verified our solution.

Let us look at another example.

Example 3:

Equation 1: **$4y + 3x = 15$**

Equation 2: **$^2/_3y + x = 3$**

It is not obvious whether the addition strategy or the subtraction strategy will be more helpful in this case. However, when we look more closely, we realize that by multiplying both the left side and the right side of **Equation 2** by **3**, we get a new **Equation 2** whose left side has the term (**+ 3x**). Since this term is also in **Equation 1**, we can use the subtraction strategy just as we did in the previous example. (Remember that we can modify either equation to suit our needs, as long as we do the same thing to both the left and the right of that equation.)

- We multiply both sides of **Equation 2** by **3**:

$3(^2/_3y + x) = 3 \times 3$.

$2y + 3x = 9$; this is the new **Equation 2**.

- We subtract the **left** side of the new **Equation 2** from the left side of **Equation 1**. (We did not change **Equation 1**.)

$(4y + 3x) - (2y + 3x)$

$= 4y + \mathbf{3x} - 2y - \mathbf{3x}$

$= 2y$

- Subtract the **right** side of the new **Equation 2** from the right side of **Equation 1**:

$15 - 9 = 6$

The results of the last two steps lead to:

2**y** = 6; therefore: **y** = 3.

As we did in the first example, we substitute **y = 3** into one of the equations. You can choose either one, typically the one you believe will make the arithmetic easier. Let us use *Equation 2*:

$$4y + 3x = 15$$

$$4 \times 3 + 3x = 15$$

$$12 + 3x = 15$$

$$3x = 3; \text{ therefore: } x = 1.$$

Our solution is: **y = 3, x = 1**

Let us substitute these values into *Equations 1* & *2* to verify that the equality is maintained in both equations:

Equation 1: $4y + 3x = 15$

$$4 \times 3 + 3 \times 1 = 15$$

$$12 + 3 = 15$$

$$15 = 15$$

Equation 2: $^2/_3 y + x = 3$

$$^2/_3 \times 3 + 1 = 3$$

$$^2/_3 \times {^3/_1} + 1 = 3 \quad 2 + 1 = 3$$

$$3 = 3$$

We have verified our solution.

> *Example 4*: In this example we will use the **substitution** method. Before we discuss this procedure, let us examine at the following equation:

$$y + 2x = 6.$$

If we subtract 2x from both sides of the equation, we get:

$$y + 2x - 2x = 6 - 2x$$

$$y = 6 - 2x$$

You can see that we have manipulated the equation to get the variable **y** by itself on one side. In mathematics, we say that we have expressed **y** in terms of **x**. (We could have also expressed **x** in terms of **y**. We do what makes our calculations easier. Either way, we will arrive at the correct result.)

Let us use the substitution technique to solve the two equations we used in *Example 1*.

> Equation 1: **2x + y = 8**

Equation 2: $x - y = 4$

- Add **y** to each side of *Equation 2*:

$x - y + y = 4 + y$

$x = 4 + y$

We have expressed **x** in terms of **y**. Now, we go to the **other** equation, *Equation 1*, and replace **x** with **4 + y**.

Equation 1: $2x + y = 8$

- We substitute **(4 + y)** in place of **x**:

$2(4 + y) + y = 8$

$(8 + 2y) + y = 8$

$8 + 3y = 8$

- We subtract 8 from both sides and get:

$8 + 3y - 8 = 8 - 8$

$3y = 0$

$y = 0$; (the product of **y** and **3** is zero, so **y = 0**.)

Since **x = 4 + y**, and **y = 0** from the previous step, we have:

$x = 4 + 0 = 4$

Our solution is x = 4, y = 0.

This is the same solution we got when we used the addition strategy.

In any given situation we pick the strategy that we think will make our computations easier and get us to the solution quickest. However, any one of the approaches will lead to the correct solution, if we don't make a mistake along the way. We should not spend time fretting about which strategy to choose. We use our knowledge, play detective, use our imagination, and perform mistake-free calculations.

14.3 Review Problems

Find the solution for the following problems:

i. Two enterprising teenagers, Izzi and Mark, make spending money by planting shrubs for their neighborhood homes. Izzi charges $10 for the service call, and $5 for each shrub he plants. Mark charges a flat fee of $7.50 per shrub, no service call fee. What is the number of shrubs for which Izzi's and Mark's charges are identical?

ii.	Rental car company A charges \$55/day, unlimited mileage, for a midsize rental car. Rental company B charges \$50/day, plus \$0.20/mile for the same class of car. If a family expects to drive 230 miles on their 10-day vacation, which rental company will be cheaper for them?

iii.	Solve the simultaneous equations for **x** and **y**:
   a.	$2x - 2y = 6$
   b.	$x + 3y = 4$

iv.	Solve the simultaneous equations for **x** and **y**:

   a.	$\frac{1}{2}x + y = 7$
   b.	$3x - 3y = 6$

Chapter 15 - Exponentials

15.1 Our Focus

In this chapter, we will examine the exponential form for representing a number and discuss the use of exponentials in expressions.

15.2 The Exponential Form

The **exponential** form is short hand for writing the result of multiplying a number or expression **by itself** multiple times. For example, since **3 × 3 = 9**, the exponential form for **9** is 3^2.

The little **2** on the top right of 3^2 says we have multiplied **two 3**s; it is called the **exponent,** or the **power,** of **3**. The number that is being multiplied, in this case **3**, is called the **base**.

When we multiply a number by itself, the result of the multiplication is also called the **square** of the number. For example, when we multiply **2** by **2** we get **4**. Thus, **4** is the square of **2**. Similarly, **5 × 5 = 25**, and **25** is the square of **5**. The square of **2** and the square of **5** are written in exponential form as 2^2 and 5^2, respectively.

When we multiply a number by itself three times, we get the **cube** of the number. For example, when we multiply three **4**s, we get: **4 × 4 × 4 = 64.** The result, **64**, is the **cube** of the number **4**. We write the cube of **4** as 4^3.

The exponential 3^2 is referred to in a number of different ways:

- the square of **3**, (as we said earlier)
- **3** squared,
- **3** raised to the power **2**,
- the **second** power of **3**.

The result of multiplying three **10**s is expressed in exponent form as: 10^3. The form 10^3 is similarly referred to in a number of ways:

- the cube of **10**,
- **10** cubed,
- **10** raised to the power **3**,
- the **third** power of **10**.

We don't have a popular single word for when we multiply a number by itself four or more times. For example, we refer to 5^4, the result of multiplying four **5**s, as **5 raised to the power 4,** or **the fourth power of 5**.

A variable also can be multiplied by itself many times. For example: $b \times b = b^2$, and it is referred to as **the square of b**, or **b squared**, or **b raised to the power 2,** or **the second power of b.** Multiplying ten **b**s gives b^{10}: **b raised to the power 10,** or **the tenth power of b.**

An expression also can be multiplied by itself many times, and the result can be expressed as an exponential. For example:

$$(a + 1) \times (a + 1) \times (a + 1) \times (a + 1)$$

$$= (a + 1)(a + 1)(a + 1)(a + 1)$$

$$= (a + 1)^4$$

We learned previously that: **a** × **b** = **ab**. Multiplying **ab** by itself gives:

ab × **ab**

$$= (\mathbf{ab})^2$$

$$= \mathbf{a} \times \mathbf{b} \times \mathbf{a} \times \mathbf{b}$$

$$= \mathbf{a} \times \mathbf{a} \times \mathbf{b} \times \mathbf{b}$$

$$= \mathbf{a}^2 \times \mathbf{b}^2 = \mathbf{a}^2\mathbf{b}^2.$$

We can write the result as $(\mathbf{ab})^2$ or $\mathbf{a}^2\mathbf{b}^2$.

When we **multiply** one exponential of a number, variable, or expression with another exponential of the **same** number, variable, or expression, we simply **add** the exponents. For example:

$$\mathbf{a}^2 \times \mathbf{a}^3$$

$$= (\mathbf{a} \times \mathbf{a}) \times (\mathbf{a} \times \mathbf{a} \times \mathbf{a})$$

$$= \mathbf{a}^5, \text{ which results from: } \mathbf{a}^{2+3}$$

Similarly, when we **divide** one exponential of a number, variable, or expression with another exponential of the same number, variable, or expression, we simply **subtract** the exponents. For example:

$$\mathbf{a}^5 \div \mathbf{a}^2$$

$$= \frac{\mathbf{a} \times \mathbf{a} \times \mathbf{a} \times \mathbf{a} \times \mathbf{a}}{\mathbf{a} \times \mathbf{a}}$$

$$= \mathbf{a}^3, \text{ which results from: } \mathbf{a}^{5-2}$$

A number, variable, or expression that is not written with an exponent actually has the exponent of **1**. The variable **a** has an exponent of **1**. This means that $\mathbf{a}^1 = \mathbf{a}$; that is, **a** raised to the power **1** equals **a**; similarly, $\mathbf{15}^1 = \mathbf{15}$, and

$(\mathbf{a} + \mathbf{1}) = (\mathbf{a} + \mathbf{1})^1$. It is usual not to explicitly write the exponent (**1**).

Any number, variable or expression raised to the power **zero** equals **1**. Thus, $\mathbf{a}^0 = \mathbf{1}$; similarly, $\mathbf{15}^0 = \mathbf{1}$, and $(\mathbf{a} + \mathbf{1})^0 = 1$.

Again, exponentials are nothing more than a short hand way of writing repeated multiplication.

A base and its exponent constitute a single number. We must **not** separate the base from the exponent, perform an operation on the base, then re-apply the exponent. For example, in the following expression: 5×3^2, we must first convert the exponential to **9**, then do the multiplication by **5** to get **45**.

Similarly, in the expression:

$$a(b + c)^2,$$

the exponent **2** applies strictly to the grouped part. Before we can multiply by **a**, we must first take the square of **(b + c)**.

For example, given **a = 2**, **b = 3**, **c = 4**, we first evaluate **(b + c)** to get **7**, take the square of **7** to get **49**, then multiply **2** and **49** to get **98**.

15.3 Negative Exponents

We learned above that a^1 = **a.** We know from **Fractions** that we can write **a** as $^a/_1$, and the reciprocal of $^a/_1$ is $^1/_a$. We write $^1/_a$ in exponential form as a^{-1}. That is to say, a^{-1} is the reciprocal of a^1, and vice versa. The product of any number or expression and its reciprocal is **1**, therefore:

$$a \times a^{-1} = a^1 \times a^{-1} = 1.$$

This is consistent with the fact that the sum of the exponents is zero.

The reciprocal of the exponential a^x is:

$$\frac{1}{a^x}$$

We write it in exponential form as: a^{-x}.

The base **a** is any number or expression and **x** is any exponent.

15.4 Simplifying a Variable Expression - Revisited

We previously discussed the simplification of variable expressions. Let us apply the knowledge we gained to an expression involving variables that have exponents.

   Example 1: $a^2(a + b - 2) + 7a^2$

We multiply a^2 through the grouped part and get:

   $a^3 + a^2b - 2a^2 + 7a^2$

We observe that $2a^2$ and $7a^2$ are common terms because they have the same variable **a** raised to the same exponent: **2**. The term a^3 has the variable **a**, but its exponent is **3**. Thus, it is not common with the other **a** terms. And, the term a^2b is a product of a^2 and **b**, and thus is not common with the other terms.

We consolidate the common terms:

   $-2a^2 + 7a^2 = 5a^2$

The final result is: $a^3 + a^2b + 5a^2$.

Remember that common terms must have the same variable or product of variables, and in the same exponents.

Let us consider another example.

> ***Example 2***: a(a + 2)
>
>   = aa + 2a
>
>   = a^2 + 2a

The expression a^2 + 2a is the exponential form of the expression: a(a + 2). It results from the use of the Distributive Property. The initial form a(a + 2) has been written as the product of the factors a and (a + 2). We will encounter many examples of both forms as we discuss subsequent topics.

15.5 Evaluation of a Variable Expression – Revisited

Now that we are familiar with exponents, let us evaluate a variable expression that involves exponents.

> ***Example 1***: Evaluate the following expression:
>
> $(ab - bc) \div (c^2 - a^2)$, given a = 3, b = 2, c = 5.
>
> $(ab - bc) \div (c^2 - a^2)$ is also written as:
>
> $$\frac{ab - bc}{c^2 - a^2}$$

This form makes it easier to see which part is the numerator and which part is the denominator.

We replace the variables in the expression with their assigned numeric values, accounting for the exponents.

$$\frac{ab - bc}{c^2 - a^2} \text{ becomes: } \frac{3 \times 2 - 2 \times 5}{5 \times 5 - 3 \times 3}$$

$$= \frac{6 - 10}{25 - 9}$$

$$= \frac{-4}{16} = -\frac{1}{4}$$

Let us work on another example.

> ***Example 2***: Evaluate the following:
>
> $$\frac{2ab + bc}{c^3 - a^2}$$

given $a = {}^1/_2$, $b = 2$, $c = {}^2/_3$.

Again, we replace the variables in the expression with their assigned numeric values, accounting for the exponents.

Let us get the value of the dividend (or numerator) first, then do the same for the divisor (or denominator).

The numerator expression **2ab + bc** becomes:

$$(2 \times {}^1/_2 \times 2) + (2 \times {}^2/_3)$$

$$= 2 + {}^4/_3$$

$$= {}^6/_3 + {}^4/_3$$

$$= {}^{10}/_3$$

The denominator expression **$c^3 - a^2$** becomes:

$$({}^2/_3 \times {}^2/_3 \times {}^2/_3) - ({}^1/_2 \times {}^1/_2)$$

$$= {}^8/_{27} - {}^1/_4 = {}^5/_{108}$$

Now that we have intermediate values for the numerator and the denominator, we can complete the evaluation of the expression by dividing the numerator by the denominator:

$$ {}^{10}/_3 \div {}^5/_{108} $$

$$ = {}^{10}/_3 \times {}^{108}/_5; \text{ (result of taking reciprocal of divisor and changing} $$

division to multiplication.)

$$ = 2 \times 36 = 72 $$

15.6 Review Problems

Evaluate the following expressions given $a = 5$; $b = 4$, $c = {}^1/_2$.

i. $a^3 + 4a + 36 =$

ii. $4a^2 \div b + b^2 + a + b =$

iii. $(a^3 - 2a^2 + 4a + 3) \div (b^2 - a^2) =$

iv. $1 \div c^2 =$

v.   A man's age is 4 times the age of his son, and 4 years less than the **square of the age** of his daughter. If his son is 8, how old is his daughter?

vi.   A store packaged 5 cases of a rare brand of wine for sale. Each case contained the same number of bottles. The total sale price was $1440, which is 10 times the square of the number of bottles in a case, expressed in dollars. What was the price per bottle?

Chapter 16 - Square and Cube Roots

16.1 Our Focus

In this chapter, we will use the knowledge we gained in our discussion of exponentials to examine square and cube roots and familiarize ourselves with the symbols for roots.

16.2 What are Roots of Numbers?

We learned earlier that when we multiply a number, variable, or expression by itself, we get a second order result or a square. We learned, for example, that the square of **2** is **4**, the square of **4** is **16**, the square of **a** is a^2, the square of a^2 is a^4, and so on.

Sometimes we have to work backwards; we are given a number, say **4**, and we have to determine the number which when multiplied with itself gives **4**. Clearly the answer is **2**, in this case. In this relationship, **2** is the **square root** of **4**. Similarly, **a** is the square root of a^2 because **a** multiplies with itself to give a^2.

The symbol for square root is: $\sqrt{}$. If we want to find the square root of **16**, we write **√16**. We know that **4** × **4** = **16**, so **√16 = 4**. But is this the only answer? No! When we multiply (**-4**) by (**-4**), the result is also **16**. (Remember this from our discussion of negative numbers?) Therefore **√16** is either **4** or (**-4**). We write this as **√16** = **±4**. The symbol (**±**) stands for: **positive** or **negative** (or, **plus** or **minus**). We say that the square root of **16** is positive or negative **4**.

Every positive number, variable, or expression has a positive and a negative square root.

We will not discuss the square roots of negative numbers. They belong to more advanced mathematics.

We learned earlier that when we multiply a number by itself **3** times we get the cube of the number. For example, **27** is the cube of **3**. It follows that **3** is the **cube root** of **27**. Similarly, **2** is the cube root of **8**. The cube root of a positive number is positive, because only positive numbers multiply **3** times and yield a positive result.

When a negative number is multiplied **3** times, the result is negative. For example:

> (**-2**)(**-2**)(**-2**) = (**-8**).

We multiply the first two negative numbers and get a positive (**+4**), then we multiply (**+4**) with the third negative number and get a negative (**-8**). Consequently, the cube root of a negative number is negative.

The symbol for **cube root** is: $\sqrt[3]{}$, thus: $\sqrt[3]{8}$ = **2**, and $\sqrt[3]{-8}$ = **-2**.

16.3 Roots as Exponents

We can express roots using the exponential form. In this case, the exponent is a fraction. An exponent of $\frac{1}{2}$ means square root. An exponent of $\frac{1}{3}$ means cube root, and so on. This alternative way of indicating roots makes the following equivalent. As usual, **a** can be any number.

$a^{\frac{1}{2}}$ is equivalent to $\sqrt{a}$, and

$a^{\frac{1}{3}}$ is equivalent to $\sqrt[3]{a}$, and so on.

Let us look at some basic roots:

- $\sqrt{100} = \pm10$; because $10 \times 10 = 100$, and $(-10) \times (-10) = 100$.

- $\sqrt{144} = \pm12$; because $12 \times 12 = 144$, and $(-12) \times (-12) = 144$

- $\sqrt[3]{512} = 8$; because $8 \times 8 \times 8 = 512$.

- $\sqrt[3]{-512} = -8$; because $-8 \times -8 \times -8 = -512$.

- $\sqrt{b^2} = \pm b$; because $b \times b = b^2$, and $(-b) \times (-b) = b^2$.

- $64^{\frac{1}{2}} = \sqrt{64} = \pm8$; because $8 \times 8 = 64$, and $(-8) \times (-8) = 64$.

- $1728^{\frac{1}{3}} = \sqrt[3]{1728} = 12$; because $12 \times 12 \times 12 = 1728$.

- $(-1728)^{\frac{1}{3}} = \sqrt[3]{-1728} = -12$; because $-12 \times -12 \times -12 = -1728$.

Now, let us work on some examples:

- Evaluate: $\sqrt[3]{(a^3 - b^2 - c)}$; given $a = 8$, $b = 20$, $c = 48$.

  First, we evaluate the grouped part: $a^3 - b^2 - c$

  $= 8^3 - 20^2 - 48$

  $= 512 - 400 - 48$

  $= 64$

  Then we take the cube root of the result: $\sqrt[3]{64} = 4$.

- Evaluate: $\sqrt{(b^2 - 4ac)}$; given: $a = 2$, $b = 5$, $c = 2$.

  First, we evaluate the grouped part: $b^2 - 4ac$

  $= 5^2 - 4 \times 2 \times 2$

  $= 25 - 16$

  $= 9$

  Therefore: $\sqrt{(b^2 - 4ac)} = \sqrt{9} = \pm3$.

A good calculator has a square root function, and possibly a cube root function, so, when in doubt, use a calculator. However, unless it is a scientific calculator, it may not show the negative of a square root. It may give half an answer.

16.4 Review Problems

Solve the following problems:

i. $\sqrt{(b^2 - 4ac)}$; given: a = 2, b = 5, c = 2.

ii. $1 - \sqrt{(b^2 - 4ac)}$; given: a = 1, b = 8, c = 7

iii. $1 + \sqrt{(b^2 - 4ac)}$; given: a = 1, b = 8, c = 7

iv. $\sqrt{169}$ =

v. $\sqrt[3]{-125}$ =

vi. $\sqrt{1024}$ =

vii. $\sqrt{356}$ =   (Use a calculator; round to 3 decimal places.)

Chapter 17 - Polynomials

17.1 Our Focus

In this chapter, we will discuss the structure of a polynomial. We will learn how to create a polynomial and how to factor a polynomial.

17.2 The Structure of a Polynomial

A **polynomial** is an expression consisting of one or more variables and possessing the following properties:

- Its terms may have **coefficients** and **positive** integer exponents.
- Its operations involve only **addition**, **subtraction**, and **multiplication**.

This sounds like a mouthful, so we will start with an example.

Example 1: $x^2 - 2x + 8$

Example 1 is a polynomial. It has one variable: **x**, and terms that include the first and second exponents of **x** and the numeral **8**. The numeral is referred to as a **numeric constant**. The x^2 term has the coefficient (**1**) and the **x** term has coefficient (-2). **Coefficient** stands for **numeric multiplier**.

Here is another example.

Example 2: $x^2 - 2xy + 3y^2 - 11$.

Example 2 is also a polynomial. It has two different variables: **x, y**. The x^2 term has coefficient (**1**), the **xy** term has coefficient (-**2**), and the y^2 term has coefficient (**3**). It has the numeric constant (-**11**).

We will only work with single variable polynomials like **Example 1**.

The word polynomial has two parts: **poly** and **nomial**. **Poly** means **many** and **nomial** means **term**. **Example 1** has three terms: x^2, **-2x**, and **8**, so we refer to it as a **Trinomial**; **Tri** means **3**. A polynomial with only one term is referred to as a **Mono**-nomial, and a polynomial with two terms is referred to as a **Bi**nomial.

These are important words to know and understand, but we don't need to fret about them. Generally, we use the word **polynomial** no matter how many terms.

In the polynomial: $x^2 - 2x + 8$, the highest power, or exponent, of the variable **x** is **2**. Therefore, it is a **second order** polynomial of **x**. The phrase **second order** refers to the highest exponent of **x** in the polynomial, which is **2**. Here is another polynomial: $x^3 - 3x^2 + x + 10$. In this example, the highest exponent of **x** is **3**, so it is a **third order** polynomial of **x**. Here is yet another polynomial: $x^4 - x$. In this last example the highest exponent of **x** is **4**. It is a fourth order polynomial of **x**. You can see how it goes.

In the discussion of the two previous examples we mentioned variables and coefficients. We already know what variables are, but what are coefficients? Let us look at a **general** second order trinomial (has three terms):

$$ax^2 + bx + c$$

The letters: **a**, **b** are place holders for integer multipliers in the terms that contain the variable **x**. They are **coefficients** of the variable terms; the coefficient of the x^2 term is **a**, and the coefficient of the **x** term is **b**. The term: **c** is a standalone integer; it is not multiplying a variable. It is called a **numeric constant**. We can also think of **c** as the coefficient of the term x^0. Remember that any number, variable, or expression raised to the power zero equals **1**. So, $c \times x^0 = c \times 1 = c$.

Let us map **a**, **b**, and **c** to the polynomial: $x^2 - 2x + 8$. The coefficient of x^2 is **1**. (Remember: $1 \times x^2 = x^2$.) The coefficient of **x** is **-2**; negative because the term is being subtracted. The number **8** is the numeric constant. In this mapping, **a** = **1**, **b** = **-2**, and **c** = **8**. *It is important to know and understand how to do this mapping!*

Polynomials result from arithmetic operations involving expressions. Let us create a second order polynomial by multiplying two first order expressions. I hope you have caught on to what **first order** means: the highest exponent of the variable is **1**.

Example 3: Create a polynomial by multiplying the following two expressions:

Expression 1: x

Expression 2: x − 2

- We multiply the two expressions:

  $x \times (x - 2)$

  $= x(x - 2)$.

- We multiply **x** through (**x − 2**) and get:

  $(x \times x - x \times 2)$

  $= (x \times x) - (x \times 2)$; because we do multiplication before subtraction.

  $= x^2 - 2x$; because $x \times x = x^2$, and $x \times 2 = 2x$.

In the resulting polynomial, the coefficient of the x^2 term is **1**, the coefficient of the **x** term is **-2**. The numeric constant is zero so it does not appear. That makes our result a binomial, but we will continue to refer to it as a polynomial.

The two expressions we started with: **x** and **x − 2**, are **factors** of the polynomial $x^2 - 2x$ because $x^2 - 2x$ is the product of **x** and **x − 2**, the same way that **4** and **5** are factors of **20** because **20** is the product of **4** and **5**.

A polynomial is the product of its factors. Conversely, factoring a **second** order polynomial produces **two** first order expressions; factoring a **third** order polynomial produces **three** first order expressions, or a first order and a second order expression, and so on.

In mathematics, we often need to find the factors of a polynomial, so Let us work on a simple example.

Example 4: Find the factors of the polynomial $2x^2 + 5x$.

This polynomial is similar to the one we created in *Example 3*. Inspecting the polynomial term by term, we notice that **x** is a common factor in both terms.

Let us return to the Distributive Property. It says:

$$ab + ac = a(b + c); \quad a, b, c \text{ may be any variables, numerals, or expressions.}$$

In the Distributive Property statement, **a** is a common factor of the terms on the left side of the equal sign. When we factor **a** out, we re-write the expression as the product of the factor **a** and the factor (**b + c**), as shown on the right side.

Similarly, in the polynomial $2x^2 + 5x$, the variable **x** is a common factor of the terms $2x^2$ and 5**x**, so we factor out **x** and re-write the expression as:

$$x(2x + 5).$$

Therefore, $2x^2 + 5x$ is the product of **x** and **2x + 5**. Conversely, **x** and **2x + 5** are the factors of $2x^2 + 5x$. (***Question***: What are the coefficients of the x^2 term and the **x** term in the polynomial: $2x^2 + 5x$?)

17.3 Factoring a Polynomial - A Closer Look

Polynomials are an important part of mathematics. Understanding how to factor a polynomial is therefore an important skill. So, let us create another polynomial and use it to further examine the process of factoring a polynomial.

> ***Example 1***: Create a polynomial by multiplying the following 2 expressions:
>
> Expression 1: x + 2
>
> Expression 2: x – 3

The polynomial we want is given by:

$$(x + 2) \times (x - 3).$$

We usually write this as:

$$(x + 2)(x - 3).$$

The color highlights are aids for subsequent steps.

Here is how we go about performing the multiplication.

- We take **x** from the **left** group and multiply it through the right group. The result is:

  $$x(x - 3) = x^2 - 3x$$

- Now, we take **2** from the **left** group and multiply it through the right group. The result is:
  $$2(x - 3) = 2x - 6$$

- Since the operation in the **left** group is addition (+), we add the results of the two steps. We get:

  $$(x^2 - 3x) + (2x - 6)$$

$$= x^2 - 3x + 2x - 6$$

$$= x^2 - x - 6; \text{ this is the resulting polynomial.}$$

In this polynomial, the coefficient of x^2 is **1**, the coefficient of **x** is **-1**, and the numeric constant is **-6**. Now Let us examine how these coefficients and the numeric constant relate to the two expressions we multiplied to get the polynomial. Again, the two expressions are:

$(x + 2)$ and $(x - 3)$.

Observation 1: The numeric constant (**-6**) in the polynomial is the **product** of the numeric constants: **2** and **-3** in the factors.

Observation 2: The coefficient of the **x** term in the polynomial is the **sum** of the numeric constants in the factors: **2 + (-3) = -1**.

Observation 3: The x^2 term, is the product of the **x** terms in the factors. Its coefficient is **1** because the **x** term in each factor has coefficient **1**.

Here is the conclusion we draw from these observations:

If the coefficient of the x^2 term of a second order polynomial is **1**, then the factors of the polynomial have the general form $(x + a)(x + b)$. The **x** terms in the factors have coefficient **1**, and **a** and **b** are unknown numbers. Factoring the polynomial involves determining the values of **a** and **b**.

Let us multiply the two general factors and see what the resulting general polynomial looks like:

$(x + a)(x + b)$

- We take **x** from the first factor and multiply it through the second factor. The result is: $x(x + b) = x^2 + bx$

- Now, we take **a** from the first factor and multiply it through the second factor. The result is: $a(x + b) = ax + ab$

- Since the operation in the first factor is addition (+), we add the results of the previous two steps. We get:

$(x^2 + bx) + (ax + ab)$

$= x^2 + bx + ax + ab; \text{ the } \textbf{x} \text{ terms are common.}$

$= x2 + x(b + a) + ab$

The product **ab** of the constants **a** and **b** produces the numeric constant in the polynomial. The sum (**b + a**) of the constants **a** and **b** produces the coefficient of the **x** term.

The factors for the polynomial: $x^2 - x - 6$, are $(x + 2)$ and $(x - 3)$, as we saw from our previous example. Mapping $(x + 2)(x - 3)$ to the general form $(x + a)(x + b)$, the numeric constant **a** = 2, and **b** = (-3). Let us confirm this by going back to our observations.

- From Observation 1: **ab** = **2** × **-3** = **-6**. (Numeric constant in polynomial.)

- From Observation 2: $a + b = 2 + -3 = -1$. (Coefficient of **x** term in polynomial.)

Let us apply this insight to factor another polynomial.

Example 2: Factor the following polynomial:

$x^2 + 7x + 12$

We know that the factors have the form $(x + a)(x + b)$ because the coefficient of the x^2 term is **1**.

From Observation 1: **ab** = 12.

From Observation 2: **a + b** = 7.

We ask ourselves: What are the two numbers whose product is **12** and whose sum is **7**? Clearly, the answer is: **3** and **4**.

The factors of the polynomial are $(x + 3)$ and $(x + 4)$.

Let us verify our result by multiplying the factors: $(x + 3)(x + 4)$.

- Multiply $(x + 4)$ by **x** from the first group:

  $x(x + 4)$

  $= x^2 + 4x$

- Multiply $(x + 4)$ by **3** from the first group:

  $3(x + 4)$

  $= 3x + 12$

- Add the two results (because the operator in the first group is addition):

  $x^2 + 4x + 3x + 12$

  $= x^2 + 7x + 12$; this is the polynomial we started with.

What happens when either **a** or **b** is zero in the factors $(x + a)$ and $(x + b)$?

Since the numeric constant in the polynomial is the product of **a** and **b**, if either **a** or **b** is zero, the numeric constant in the polynomial will be zero. For example, the factors for the polynomial: $x^2 - 2x$, from *Example 1*, are: (x) and $(x - 2)$. We have written the first factor as (x) instead of $(x + 0)$.

What happens when both **a** and **b** are zero?

- The factors will be $(x + 0)$ and $(x + 0)$.

- The numeric constant will be zero because **ab** will equal **0**.
- The **x** term will also be zero because its coefficient, given by $(a + b)$ will be zero. Remember that a coefficient is a multiplier, and multiplying anything by zero gives zero.

The polynomial will be reduced to the mono-nomial: x^2.

Let us look at another example of factoring.

Example 3: Factor the following polynomial:

$3x^2 - 11x + 6$

In this example, the x^2 term has a coefficient of **3**. This means that the factors will have the form (**3x + a**)(**x + b**).

Note that the **x** term in one of the **factors** has to have a coefficient of **3** for the polynomial to have x^2 term coefficient of **3**. If, for example, the x^2 has the coefficient **6**, the factors would have the form:

(**x + a**)(**6x + b**) or (**2x + a**)(**3x + b**).

In order words, the coefficients of the **x** terms in the factors of the polynomial are themselves factors of the coefficient of the x^2 term in the polynomial. We take this into consideration when we factor a polynomial.

Let us multiply the general factors and compare the result with the polynomial.

(3x + a)(x + b)

= 3x(x + b) + a(x + b)

= $3x^2$ + 3xb + ax + ab

= $3x^2$ + x(3b + a) + ab

Comparing the polynomial:

$3x^2 - 11x + 6$

with the result:

$3x^2$ + x(3b + a) + ab,

we determine the following:

- **ab** = 6
- 3b + a = -11

First, the product of **a** and **b** is **6**. So, we are looking for two numbers, **a** and **b**, which are **factors of 6**. Next, multiplying one of those **factors** by **3** and adding the other factor, gives: (**-11**). This is a hint that we are looking for two factors of **6** that are both negative. (Remember that the product of two negatives is positive, so this is consistent with: **ab** = **6**, and the sum of two negatives is negative, which is consistent with: 3b + a = **-11**.)

Now, let us do some detective work. We start with a = **-3**, b = **-2**, which are factors of **6**.

- The product: **ab** is **6**.
- The **x** term coefficient (**3b + a**) gives:
  (3 × -2 + -3)

$$= -6 - 3$$

$$= -9 \neq -11; \ (\text{-9 is not equal to -11})$$

We can see that **a** = **-3**, **b** = **-2** doesn't give us the factors we are looking for.

Let us try **a** = **-2**, **b**= **-3**.

- The product: **ab** = **6**.
- The **x** term coefficient (**3b + a**) gives:

$$(3 \times \text{-3} + \text{-2})$$

$$= \text{-9} - 2$$

$$= \text{-11}$$

We have matched the polynomial's x^2 coefficient of **3**, **x** coefficient of **-11**, and the numeric constant **6** with the values: **a** = **-2** and **b**= **-3**.

We substitute **a** = **-2** and **b**= **-3** into the general factors (**3x + a**)(**x + b**) and get the solution:

(**3x − 2**)(**x − 3**); subtraction because **a** and **b** are both negative.

You can see that sometimes we must go through an investigative process and a little trial and error. It helps to call on our accumulated mathematical insights, especially our knowledge of factoring and of simplifying variable expressions. What we should not do is random guessing.

In the process of finding the factors of the polynomial in Example 3, we encountered the following two simultaneous equations:

i.   **ab** = **6**

ii.   **3b + a** = **-11**

Why did we not use the techniques we learned in solving simultaneous linear equations to find values for **a** and **b**? The answer is that in that discussion we only considered **linear** equations. The first of the two equations above is non-linear, so the techniques we learned previously do not apply.

17.4 Review Problems

i.   Create polynomials by completing each multiplication.

- (x − 2)(x +1)

- (2x + 1)(3x + 5)

- (2x − 1)(3x − 5)

- (2x + 1)(3x − 5)

ii. Factor the following polynomials.

1. $x^2 - 2x + 1$

2. $x^2 - x$

3. $4x^2 - 18x - 10$

Chapter 18 - Solution of a Quadratic Equation

18.1 Our Focus

In this chapter, we will examine how a quadratic equation relates to a polynomial and discuss the process for solving a quadratic equation.

18.2 The Quadratic Equation

An equation of the form: $ax^2 + bx + c = 0$ is called a quadratic equation. It is an equation because it declares an equality. The word **quadratic** comes from the Latin for square, which relates to the fact that the expression on the left is a second order polynomial.

The above polynomial is in general form. We recognize it from the earlier discussion of polynomials. The letters **a** and **b** are the coefficients of the x^2 and **x** terms, respectively, and **c** is a numeric constant, just like we saw before. Let us familiarize ourselves again with the values that correspond to **a**, **b**, **c** by examining the following specific polynomials:

1. Polynomial: $x^2 - 2x + 1$

    a = 1, **b** = -2, **c** = 1

2. Polynomial: $x^2 - 1$

    a = 1, **b** = 0 (0 × **x** = 0), **c** = -1

3. Polynomial: $4x^2 - 18x - 10$

    a = 4, **b** = -18, **c** = -10

When we solve a quadratic equation, we find the values of **x** for which the polynomial evaluates to zero, so that the equality is maintained. If a solution exists, there are always two values of **x** that make this so. Sometimes the values are the same. We solve the equation by following these steps:

- We factor the polynomial.

- We set each factor equal to zero.

- We solve for **x** in each factor.

Let us look at an example.

>   ***Example 1***: Solve the following quadratic equation:
>
> $x^2 - 3x + 2 = 0$

In order to find the solution, we first factor the polynomial using the techniques we learned earlier:

- We factor the polynomial:

  $$x^2 - 3x + 2 = (x - 1)(x - 2)$$

- We find the values of **x** that make the polynomial equal **0**:

  Setting: **x − 1 = 0,** gives: **x = 1**.

  Setting: **x − 2 = 0**, gives: **x = 2**.

  The solution for the equation: $x^2 - 3x + 2 = 0$ is:

  x = 1, x = 2.

When we substitute x = 1 into the polynomial, we get:

$$1^2 - 3 \times 1 + 2$$

$$= 1 - 3 + 2$$

$$= 0$$

Similarly, when we substitute **x = 2** into the polynomial, we get:

$$2^2 - 3 \times 2 + 2$$

$$= 4 - 6 + 2 = 0$$

We have verified equality with both **x = 1, x = 2.**

The significance of solving a quadratic equation will become apparent when we discuss graphs in the Cartesian coordinate system. For now, let us just accept that it is important to know how to solve quadratic equations.

Let us look at another example:

Example 2: $4x^2 - 18x - 10 = 0$

- We factor: $4x^2 - 18x - 10$, and get: (4x + 2)(x - 5)

- Therefore: $4x^2 - 18x - 10 = (4x + 2)(x - 5) = 0$

- Setting: 4x + 2 = 0, gives x = $-^1/_2$.

- Setting: x − 5 = 0, gives x = 5.

  The solution is: **x** = $-^1/_2$, **x** = 5.

Again, we go through our verification; first with **x** = $-^1/_2$:

$$4x^2 - 18x - 10$$

$$= 4 \times (-^1/_2)^2 - 18 \times (-^1/_2) - 10$$

$$= 4 \times {}^1/_4 - (-^{18}/_2) - 10$$

$$= 1 + 9 - 10 = 0$$

Next, with **x = 5**:

$$4x^2 - 18x - 10$$

$$= 4 \times 5^2 - 18 \times 5 - 10$$

$$= 100 - 90 - 10 = 0$$

We have verified the solution: **x = $-^1/_2$, x = 5** by confirming equality.

18.3 Review Problems

Solve the following quadratic equations:

i. $x^2 - 4 = 0$

ii. $2x^2 + 7x - 4 = 0$

iii. $4x^2 + 6x - 4 = 0$

iv. $x^2 - x - 42 = 0$

Chapter 19 - The Quadratic Formula

19.1 Our Focus

With a little experience, we can usually factor the polynomial in a quadratic equation and solve the equation. But there are times when we cannot easily accomplish this. The **Quadratic Formula** provides a sure way to solve a quadratic equation, if a solution exists. In this chapter, we will learn how to use the quadratic formula.

19.2 The Quadratic Formula

Here is the quadratic formula:

$$x = \frac{-b \pm \sqrt{b^2 - 4ac}}{2a}$$

a, **b**, and **c** in the formula correspond to **a**, **b** and **c** in the generalized polynomial: $ax^2 + bx + c$.

In order to solve a particular quadratic equation, we substitute the corresponding values of **a**, **b,** and **c** from the equation into the formula. We then evaluate the formula. The formula yields two values for **x**, which constitute the solution to the equation.

Note that in the formula the symbol (**±**) appears to the left of the square root symbol. It accounts for the positive and a negative square roots that we discussed. The positive root results in addition; the negative root results in subtraction. That is why we get two values for **x**.

As usual, let us look at an example.

Example 1: Solve the quadratic equation $x^2 - 3x + 2 = 0$, using the quadratic formula. This is one of the examples we used previously. We determined that the solution is:

x = 1, x = 2.

We should arrive at the same solution using the quadratic formula.

First, we determine the values that correspond to: **a**, **b**, and **c.**

a = 1, because the coefficient of x^2 term is **1**.

b = -3, because the coefficient of the x term is **-3**.

c = 2, because the numeric constant is **2**.

Next, we substitute the values into the quadratic formula, and evaluate it step by step.

The square root part in the numerator: $\sqrt{(b^2 - 4ac)}$ becomes:

$$\sqrt{[(-3)^2 - 4 \times 1 \times 2]}$$

= √[9 − 8] = ±1; (because 1 × 1 = 1 and -1 × -1 = 1).

The (**±**) in the formula tells us to evaluate using one root, then the second root. That is how we arrive at two solutions.

We are still on the numerator part of the formula.

We replace the square root part with the result from the previous step. We get:

-**b** ± **1**. Since **b** = -3, -**b** = -(-**3**), which is +**3**, or simply **3**.

-**b** ± **1** becomes: **3** ± **1**, which is: **3** + **1** = **4**, and **3** − **1** = **2**.

The numerator has yielded two values, **4** and **2**.

Now, we evaluate the denominator: 2**a**.

Since **a** = 1, 2**a** = 2 × 1 = 2.

Finally, we complete the evaluation of the quadratic formula by dividing each of the two values we got for the numerator by the result we got for the denominator:

4 ÷ 2 = 2, and 2 ÷ 2 = 1.

Therefore, the solution for the quadratic equation: $x^2 − 3x + 2 = 0$, is:

x = 1, **x** = 2.

Points to remember:

- We must correctly assign values to **a**, **b**, and **c** in the formula.
- Subtraction makes a coefficient or a numeric constant negative.
- We must correctly apply the rules for operations involving positive and negative numbers. For example, if the corresponding value of **b** is negative, then (-**b**) in the formula becomes positive.
- (±) in the formula tells us to apply both the positive and negative square roots.
- Applying both roots gives us two solutions.

When we do these things, the quadratic formula is easy to use and a great help!

Let us work on one more example.

Example 2: Solve the quadratic equation $4x^2 − 18x − 10 = 0$, using the quadratic formula.

This is another one of the examples we used previously. We determined then that the solution is: **x** = -$^1/_2$ or **x** = 5. Again, we should arrive at the same solution using the quadratic formula.

- We start by first determining the values that correspond to **a**, **b**, and **c**.
  a = 4, because the coefficient of x^2 term is 4.

  b = -18, because the coefficient of the x term is -18.

  c = -10, because the numeric constant is -10.

- We substitute the values into the quadratic formula, and evaluate it step by step. The square root part in the numerator: $\sqrt{(b^2 - 4ac)}$ becomes:

$\sqrt{[(-18)^2 - 4 \times 4 \times (-10)]}$

$4 \times 4 \times (-10) = -160$, and $- (-160) = + 160$.

Therefore: $\sqrt{[(-18)^2 - 4 \times 4 \times (-10)]}$

$= \sqrt{(324 + 160)}$

$= \sqrt{484} = \pm 22$

We are still on the numerator part of the formula.

- We replace the square root part with result from the previous step. We get:

$-b \pm 22$. Since b -18, $-b = -(-18)$, which is $+18$, or simply **18**.

$-b \pm 22$ becomes: **18** $+ 22 = 40$, and **18** $- 22 = -4$.

The numerator has yielded two values, **40** and **-4**.

Now we evaluate the denominator: 2**a**.

Since **a** $= 4$, 2**a** $= 2 \times 4 = 8$.

Now we complete the evaluation of the quadratic formula by dividing each of the two results we got from the numerator by the result we got from the denominator:

$40 \div 8 = 5$, and $-4 \div 8 = -\frac{1}{2}$.

The solution for the quadratic equation: $4x^2 - 18x - 10 = 0$, is:

x = 5, x = -$\frac{1}{2}$.

19.3 Review Problems

Solve the following quadratic equations, first by inspection and factoring, then by using the quadratic formula.

   i. $x^2 - 4 = 0$

   ii. $2x^2 + 7x - 4 = 0$

   iii. $4x^2 + 6x - 4 = 0$

   iv. $x^2 - x - 42 = 0$

Chapter 20 – Basic Geometry

20.1 Our Focus

Our discussion of basic geometry will be focused on the properties of the straight line, the triangle, the rectangle (of which the square is a special case), the circle, and the cylinder.

20.2 The Straight Line

Figure 1: Lines and Angles

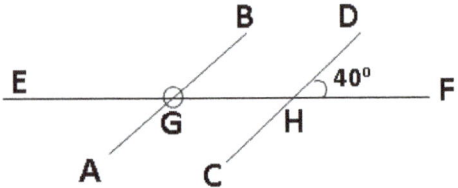

A line consists of points, as we learned earlier. A straight line is simply an infinite number of points strung together end-to-end. A straight line resides in one dimension. This means that if we follow along it, we go in one direction only. We cannot turn left or right onto the same straight line.

In geometry, we label a straight line with letters of the alphabet, so that we can distinguish one straight line from another. *Figure 1* shows three straight lines: the line AB, the line CD, and the EF. The lines AB and CD cross the line EF at the points G and H. The points G and H are points of **intersection**. Another way to say lines cross is to say they intersect. The line segment between the points G and H is line GH.

When two lines intersect, they form angles. In the figure, one of the angles line CD forms with line EF has the value **40°**, pronounced *forty degrees*. It is the angle that is sandwiched by the line segments HD and HF. The little 'o' on top of **40°** is the symbol for **degrees**. *It does not stand for 40 raised to the power zero.*

We have placed a circle around the point where lines AB and EF intersect. If we start anywhere on the circle and walk completely around the circle back to our starting point, we describe an angle of 360°. If we go exactly halfway, we describe a 180° angle.

We name an angle with three letters. We start with the letter that labels the endpoint of one of the line segments that sandwich the angle, followed by the letter that labels the base of the angle, then the letter that labels the end point of the second line segment. In *Figure 1*, the **40°** angle is identified as ∠DHF. The symbol ∠ stands for **angle**. If we start on the line segment HF and walk around in a semi-circle across line segment HD back to line GH, we go through a 180° angle. This tells us that since ∠DHF is 40°, ∠GHD is 140°.

The angles ∠DHF and ∠GHC are mirror images of each other. They have the same value: 40°. By the same token, angle ∠GHD = ∠CHF = 140°.

Two lines that never meet, no matter how far we extend them, are said to be parallel lines. The distance between parallel lines stays the same all the way to infinity. In *Figure 1*, the lines AB and CD are parallel.

Being parallel, lines AB and CD make identical angles with line EF at the points of intersection. Therefore, ∠BGH = ∠DHF = 40°, and ∠EGB = ∠GHD = 140°.

Figure 2: Perpendicular Lines

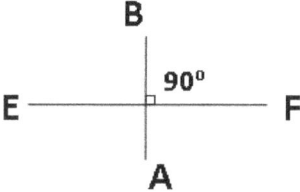

When two lines intersect in exactly east-west and north-south fashion, as shown in Figure 2, they divide the 360° angle at the point of intersection into four equal angles. The angle at each corner is 90°. The 90° angle has a special name. It is called **right angle**. In **Figure 2**, the lines EF and AB intersect at right angles.

Lines that intersect at right angles are **perpendicular** to each other. We say that they are perpendicular lines.

A right angle is usually depicted as a square corner, as shown in **Figure 2**. It is important to remember that a right angle equals 90°.

20.2.1 Review Problems

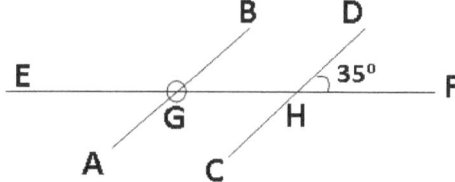

In the figure above, lines AB and CD area parallel. If ∠DHF equals = 35°,

  i. ∠EGB = _____ degrees.

  ii. ∠EGA = _____ degrees.

20.3 The Triangle

A closed shape whose sides are all straight lines is called a **polygon**. In this section, we focus on the three-sided polygon called **Triangle**. Because a triangle is bounded by three sides, it encloses three angles. That is where its name comes from. The three angles a triangle encloses add up to 180°.

Triangles can have many different shapes, and the shape gives a triangle attributes that differentiate it from other triangles. Because of this, triangles are given specific names based on their shapes. **Figure 3** shows triangles with different shapes and names. Let us take a close look at them.

Figure 3: Triangles

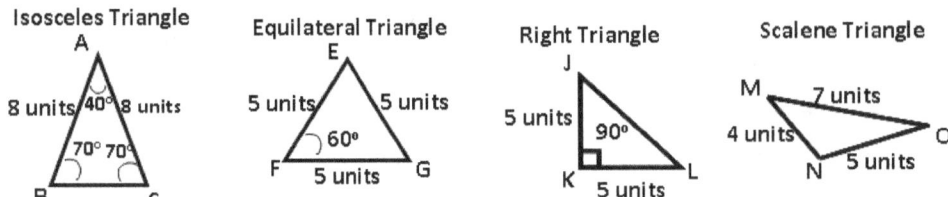

Isosceles Triangle: The first triangle in *Figure 3* is an **Isosceles** Δ. (We use the symbol Δ as short hand for the word **triangle**.) An Isosceles Δ has two sides with equal length. It gives the triangle a symmetrical shape about its middle. As such, the two angles at the base of the equal sides are also equal. In the figure, the Isosceles triangle has two sides that are both 8 units long and base angles ∠ABC and ∠ACB that are both 70°. Since the sum of the three angles is 180°, the top angle ∠BAC is 40°.

Equilateral Triangle: The next triangle in *Figure 3* is an **Equilateral** Δ. An Equilateral Δ has sides that are all equal in length. The name **Equilateral** derives from this fact. It gives the triangle a symmetrical shape about every middle. As such, each of the three angles inside the triangle equals 60°.

Right Triangle: The third triangle is a **Right** Δ. In a Right Δ, one of the three angles inside the triangle equals 90°. This means that the **sum** of the two remaining angles also equals 90°. The Right Δ in the figure happens to be isosceles because two of its sides have equal length: 5 units. However, a Right Δ need not be Isosceles. (What do you think the size of ∠JLK is?)

Scalene Triangle: The fourth triangle is a **Scalene** Δ. In a Scalene Δ, the sides have different lengths and the inside angles all have different sizes.

Obtuse Triangle: If one of the angles of a triangle is greater that 90°, the triangle is referred to as an Obtuse Δ. In *Figure 3*, we can tell by inspection that ∠MNO of the Scalene Δ is greater than 90°. That makes the triangle an Obtuse triangle as well. An obtuse triangle can be Isosceles, but it cannot be Equilateral. By definition, it can't be a Right Δ either.

The shortest distance between any two points is along the straight line that directly connects the two points. It follows that for any triangle, the sum of the lengths of any two sides is greater than the length of the third side.

20.3.1 Congruent and Similar Triangles

Two triangles whose sides have the same dimensions are called **Congruent** triangles. If we place two congruent triangles on top of each other, they align perfectly. Corresponding angles inside congruent triangles also match in size. Think of congruent triangles as *identical twin* triangles. The concept of congruence applies to other figures as well.

Figure 4: Congruent and Similar Triangles

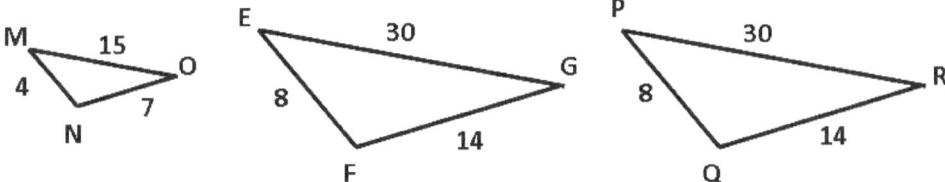

In **Figure 4**, triangles EFG and PQR are congruent.

Two triangles are similar if the **ratios** of their corresponding sides are the same. Consider triangles MNO and EFG in **Figure 4**. Each side of ΔMNO is $^1/_2$ of the corresponding side of ΔEFG. One of the two Δs is a scaled down (or scaled up) version of the other. Corresponding angles inside similar triangles match in size.

Congruent Δs are also similar because their corresponding sides all have the same ratio: **1**. However, similar Δs are not always congruent, as we can see from ΔMNO and ΔEFG.

Architects often build scaled down models of their designs. The models are miniatures or scaled down versions of the real structures that get erected from the designs. The models are **similar** to the real structures in the sense we are discussing here.

20.3.2 Perimeter and Area of a Triangle

Perimeter of a Triangle: If we start from one corner of a triangle and walk along the three sides back to our starting point, the length we cover is called the **Perimeter** of the triangle. Perimeter is simply the sum of the lengths of the three sides. It is a measure of length. For example, the perimeter of the Equilateral Δ in **Figure 3** is 15 units, because each side is 5 units long.

Area of a Triangle: The area of a triangle is the total space the triangle covers. We will use **Figure 5** to learn how to calculate the area of a Δ.

Figure 5: Area of a Triangle

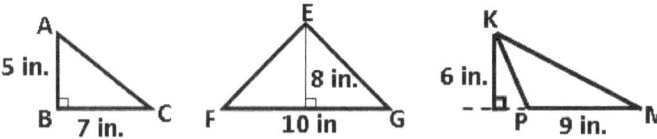

In order to find the area of a triangle, we pick one of its sides as the **base**. Then we find out the following:

i.   The **length** of the side of the triangle that we pick as the base.

ii.   The **height** of the triangle. The height of a triangle is the length of the perpendicular line from the base to the corner opposite to the base. In **Figure 5**, ΔABC is a right triangle. The side AB is perpendicular to the side BC. If we choose BC as the base, then AB is the perpendicular line from BC to the corner opposite to BC. Therefore, AB becomes the height. (If we had chosen AB as the base, BC would be the height.) For ΔEFG, we have chosen FG as the base. We have drawn a perpendicular line from FG to its opposite corner E. This becomes the height. For ΔKPM, we have chosen the side PM as the base. We have drawn a perpendicular line from an extension of PM to the opposite corner K. This line becomes the height. Note that the line that represents the height does not have to lie inside the triangle, as the third example shows.

Once we have the lengths of base and the height, we compute the area as:

$^1/_2$ × **base** × **height**.

- The area of ΔABC is: $^1/_2$ × **7** × **5** = 17.5 square ins.

- The area of ΔEFG is: $^1/_2$ × **10** × **8** = 40 square ins.

- The area of ΔKPM is: $^1/_2$ × **9** × **6** = 27 square ins.

Typically, we are given enough information to compute the base and height. Once we have done that, the area is simple to determine.

20.3.3 The Pythagorean Theorem

A right triangle has another unique property besides the right angle it encloses. The lengths of its sides bear a unique and important relationship with each other. A Greek mathematician named **Pythagoras** of **Samos** discovered the relationship the sides have with each other and formulated it into a statement known as a **theorem** in mathematics.

The theorem is appropriately named after him and is called the **Pythagorean Theorem**. Before we state the theorem, let us revisit the right triangle.

Figure 6: Right Triangle

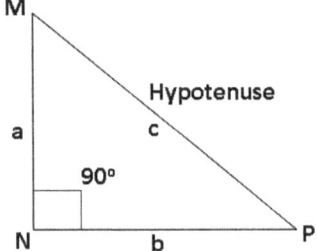

Figure 6 shows a right triangle, ΔMNP, not unlike the one we discussed previously. We have assigned the variables **a**, **b**, and **c** to the lengths of its sides. The side with length **c** is the side opposite the right angle. That side has a unique name. It is called the **Hypotenuse** of the right triangle.

The square of the length of the side MN is a^2.

The square of the length of the side NP is b^2.

The square of the length of the hypotenuse MP is c^2.

Pythagoras discovered that for every **right** triangle:

$$a^2 + b^2 = c^2$$

This is the Pythagorean theorem. It says that in a right triangle, the square of the hypotenuse equals the **sum** of the squares of the other two sides.

From our discussion of roots, we can see immediately that:

$$c = \sqrt{(a^2 + b^2)}$$

$$a = \sqrt{(c^2 - b^2)}$$

$$b = \sqrt{(c^2 - a^2)}$$

If we know the lengths of two sides of a right triangle, we use the Pythagorean theorem to calculate the length of the third side.

As an example, let us use the Pythagorean theorem to find the length **a** of the side FG of the right triangle, ΔFGL, in **Figure 7**.

Figure 7: Right Triangle - Unknown Side

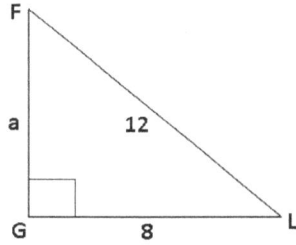

The side FL is the hypotenuse. Therefore:

$$a = \sqrt{(12^2 - 8^2)}$$

$$= \sqrt{(144 - 64)}$$

$$= \sqrt{80}$$

a = 8.944; rounded to three decimal places.

We use a calculator to determine the root of such a number.

20.3.4 Review Problems

Refer to the three Δs that follow.

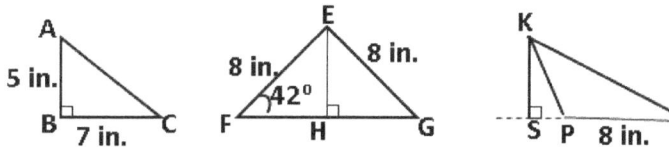

i.   The size of ∠HEG in the second triangle ΔEFG is _____ degrees.

ii. The first triangle ΔABC and the third triangle ΔKPM have equal areas. What is the length of line KS in the third triangle?

iii. What is the length of the side AC in ΔABC? Round your answer to two decimal places.

20.4 The Rectangle

Figure 8: The Rectangle

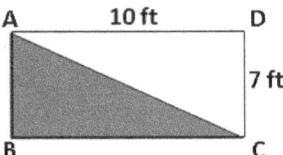

Figure 8 shows a rectangle. It is bounded by four lines, and it has right corners. Thus, ∠ABC = 90°, and so are the angles at the other corners.

A rectangle has a length and a width. The length of the rectangle in *Figure 8* is 10 feet, and its width is 7 feet. (Your rectangular living room has a length and a width. If you decide to install a carpet, you will have to know the dimensions.)

The distance from the point B to C to D to A back to B along the edge of the rectangle is called the **perimeter** of the rectangle. It is the total distance around the edge of the rectangle. The perimeter of the rectangle in *Figure 8* is **2 times the sum** of the length and width: 34 feet.

The space that the rectangle takes up is called the area of the rectangle. In the figure we have shaded **half** of this area. The area of a rectangle is **the product** of its length and its width. It is measured in **square** units. In this example, the unit is feet, so the area is 70 square feet.

The line that goes from one corner of the rectangle across the middle to the opposite corner is called the **diagonal** of the rectangle. In the figure, the line AC is a diagonal. Line DB, which is not drawn, is also a diagonal. A diagonal divides a rectangle into two equal halves. Half of a rectangle is a right triangle, and the diagonal is the hypotenuse of the right triangle.

Since the angle inside each corner of a rectangle is 90°, if we are told ∠BCA in the figure equals 30°, it means ∠ACD = 60°.

The total of the angles enclosed in the entire rectangle is: 4 × 90° = 360°.

Figure 9: The Square

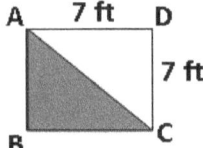

A square is a rectangle whose length equals its width. Thus, the area of a square is the **square** of its length. The area of the square in *Figure 9* is:

$7 \times 7 = 7^2 = 49$ square feet.

And, the perimeter of a square is 4 times its length. The perimeter of the square in the figure is:

$4 \times 7 = 28$ feet.

The diagonal AC is the hypotenuse of the right triangle ΔABC. We can determine the length of AC by using the Pythagorean theorem.

$AC = \sqrt{(7^2 + 7^2)}$

$= \sqrt{98} = 9.90$; rounded to two decimal places.

20.4.1 Review Problems

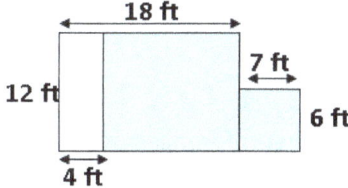

i.   Find the perimeter of the figure above.

ii.   Find the perimeter of the shaded portion.

iii. Find the area of the shaded portion.

20.5 The Circle

Figure 10: The circle

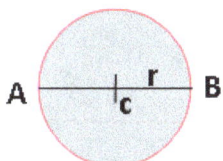

Tie one end of a piece thread to a pin and the other end to a pencil. Stick the pin into a piece of paper. Stretch the string and run the tip of the pencil completely around the pin. You trace a circle. The point where you stick the pin is the **center** of the circle. The length of the stretched thread is the **radius** of the circle.

Every point on the circle is the same distance from its center, that distance being the radius. The distance from a point on the circle through the center to the opposite point on the circle covers twice the length of the radius. It is called the **diameter** of the circle. A diameter divides the circle into two equal half-moons.

In *Figure 10*, **C** marks the center of the circle. CD is a radius of the circle. The letter **r** is usually used to represent the radius of a circle, and the letter **d** the circle's diameter. AB is a diameter. A diameter is twice the radius:

$$d = 2r.$$

The angle around the center of a circle is 360°. The angle at the center of a semi-circle (half of a circle) is 180°.

The distance around a circle is called the **circumference** of the circle. In the figure, the circumference of the circle is shown in red. Circumference of a circle is an equivalent idea to perimeter of a rectangle or a triangle.

The space inside a circle is called the area of the circle. In the figure, the area has been shaded gray.

Early Greek mathematicians discovered that the ratio of the circumference of a circle to the circle's diameter is always the same, no matter the size of the circle. They gave this ratio a special name: **Pi**. The symbol for Pi is π. **Pi** is a numeric constant whose value has proved to be impossible to determine precisely. In our problems, we will use the value $^{22}/_7$ or **3.14** as an estimate of π.

The formula for the circumference of a circle is $\pi \times d$, written as: πd. The variable **d** is the circle's diameter. We can express circumference in terms of the radius **r**. Since $d = 2r$, Circumference $= \pi d = 2\pi r$.

The formula for the area of a circle is π multiplied by the square of the radius: πr^2. Since $d = 2r$, $r = \frac{d}{2}$. Substituting $\frac{d}{2}$ for **r** in the formula for area: $\pi r^2 = \pi(\frac{d}{2})^2$.

We can determine the circumference and area of a circle in terms of diameter or radius, but we must account for the fact that $d = 2r$.

Let us work on some examples.

   Example 1: Determine the area and the circumference of a circle whose diameter is 10 inches.

We will use the formula πr^2 to compute area and the formula $2\pi r$ to compute circumference, and $\pi = 3.14$. Since diameter is **10** inches, radius is **5** inches.

- Circle's area: $\pi r^2 = 3.14 \times 25 = $ **78.5** square inches.

- Circle's circumference: $2\pi r = 2 \times 3.14 \times 5 = $ **31.4** inches. (Circumference, like perimeter, is a length and is measured in linear units.)

Example 2: Arc AB of the circle below is $^1/_4$ of the length of the circumference of the circle. **C** is the center of the circle. The circle's radius is 4 inches; $\pi = 3.14$.

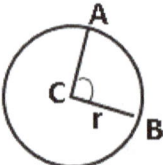

   i.   What is the size of the area bounded by the lines AC, CB, and the arc AB?

- **Solution**: The size of the area bounded by the lines AC, CB, and the arc AB has the same ratio to the total area as the arc AB has to the circumference. It is $^1/_4$ of the area of the circle. The area of the circle is:

$\pi r^2 = 3.14 \times 4^2$.

$^1/_4$ of the area is: $^1/_4 \times 3.14 \times 4^2 = 3.14 \times 4 = 12.56$ square inches.

ii. What is the value of ∠ACB in degrees?

- Solution:

  Angle ∠ACB is $^1/_4$ of the angle at center of the circle which, we

  know, is $360°$. ∠ACB = $^1/_4 \times 360° = 90°$.

20.5.1 Review Problems

Assume that $\pi = 3.14$ for the following problems.

i.   The figure above shows two concentric circles. (They have the same center.) The diameter of the bigger circle is 16 inches. The diameter of the smaller circle is $^3/_4$ of the diameter of the bigger circle. What is the area of the shaded portion?

ii. A given circle has the area 153.86 square inches.
     a) What is the radius of the circle?

     b) What is the circumference of the circle?

iii. What is the size in degrees of the angle that the hour hand of a clock circles through in 24 hours?

20.6 Surface Area and Volume

In this section we will discuss basic concepts behind the surface area and volume using two simple three-dimensional figures: **Box** and **Cylinder**.

The box is a three dimensional figure enclosed in flat rectangular surfaces. The edges at each corner of a box are orthogonal; this means the edges meet at right angles. A box has six rectangular sides. The orthogonal property ensures that two opposite sides have the same dimensions and opposite edges are parallel to each other.

As we learned earlier, a rectangle whose sides are all of equal length is called a square. A cube is a box whose edges all have the same length. In other words, a cube has square sides.

Figure 11: Box and Cube

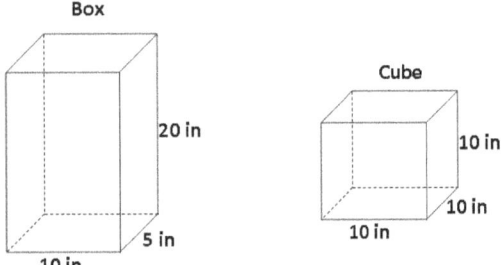

Figure 11 shows a box and a cube. Both have six sides. The **surface area** of a box is the sum of the areas of its six sides. The surface area is a measure of how much material is used to make the box, if there is no overlap. Since opposite sides of a box have the same dimensions, only three of the six sides are unique. Consequently, the surface area of a box is twice the sum of the areas of three unique sides. Here are the unique sides of the in *Figure 11*:

- Front: 20 in. by 10 in.; same as the back.
- Top: 10 in. by 5 in.; same as the bottom.
- Right side: 20 in. by 5 in.; same as the left side.

The surface area of the box is:

$$2(20 \times 10) + 2(10 \times 5) + 2(20 \times 5) \text{ square inches.}$$

$$= 400 + 100 + 200 \text{ square inches}$$

$$= 700 \text{ square inches}$$

The surface area of a cube is computed in the same way. However, things are simpler because all the surfaces have the same area. Since a cube has six surfaces like a rectangle, we multiply the area of one surface by **6** to get the cube's total surface area.

The surface area of the cube in *Figure 11* is:

$$6 \times (10 \times 10)$$

$$= 600 \text{ square inches.}$$

The **volume** of a box is a measure of its capacity, the amount of room inside it to hold stuff. We calculate the volume of a box by multiplying the area of one side by the length of the edge perpendicular to that side. In Figure 9, it can be:

- The area of the front, which is **20 × 10**, multiplied by the length **5** of the edge perpendicular to it, giving: **20 × 10 × 5 = 1000** cubic inches.

- The area of the top, which is **10 × 5**, multiplied by the length **20** of the edge perpendicular to it, giving: **10 × 5 × 20 = 1000** cubic inches.

- The area of a side, which is **20 × 5**, multiplied by the length 10 of the side perpendicular to it, giving: **20 × 5 × 10 = 1000** cubic inches.

In all three choices, the result is the same: it is the product of the box's three dimensions. Thus, the volume of a box is simply: **length × width × height**. The result is in cubic units.

Let us revisit the **units of measure** of area and volume. When we calculate area, we multiply one linear unit: **length** by another linear unit: **length**. For example, the area of the front of the box in *Figure 11* is:

20 inches × 10 inches.

The result is the product of the numbers:

20 × 10 = 200,

as well as the product of the linear units:

inches × inches = inches^2.

Therefore, the resulting unit of measure for the **area** is **square inches**.

The **volume** of a box, as we said earlier, is:

length × width × height.

Again, each dimension is in linear units. If the unit is inches, then we have:

inches × inches × inches = inches^3.

Therefore, the resulting unit of measure for the **volume** is **cubic inches**.

Figure 12: Cylinder

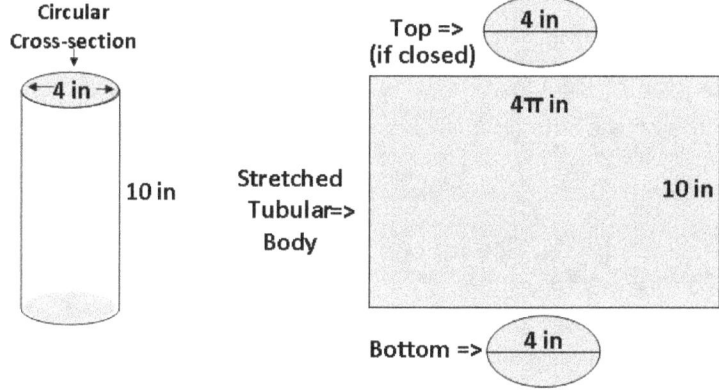

Figure 12 shows a **Cylinder**. It is essentially a circular tube. Flower vases often have a cylindrical shape with an open top. A pill container sometimes has a cylindrical shape, and is closed at both ends. A cylinder may also have both ends open; an example if the cardboard tube that paper towels are wrapped around.

The **volume** of a cylinder is the **area** of its circular cross-section **multiplied by its height**. The diameter of the circular cross-section of the cylinder in *Figure 12* is **4** inches. The radius is **2** inches, and the area of the circular cross-section is:

$\pi r^2 = \pi 2^2 = 4\pi$ square inches

Since the height of the cylinder is **10** inches, its volume is:

$4\pi \times 10 = 40\pi$ cubic inches.

Using: **π = 3.14**, the volume is:

40 × 3.14 = 125.60 cubic inches.

In order to calculate the **surface area** of the cylinder, we need to know whether both ends are open, one is closed and the other open, or both end are closed.

If both ends are open, the surface area is simply the area of the surface of the cylindrical tube. In *Figure 12*, we have slit the cylindrical tube straight down the side and stretched it out. What we get is a sheet that is **10** inches in height. Its width is exactly equal to the circumference of the circular cross-section. (Imagine folding the sheet back into a cylindrical shape.)

Circumference is **2πr**. The circumference of the circular cross-section of the cylinder in *Figure 12* is:

$2\pi r = 2\pi \times 2 = 4\pi$ inches.

Therefore, the **width** of the sheet of material the cylindrical tube is made from is: **4π** inches. The area of the sheet is:

width × **height** square inches

= **4π** × **10** square inches

= **40π** square inches

If the cylinder has open ends, its surface area is:

40π = 125.60 square inches, based on **π = 3.14**.

If it has **one** closed end, we must add the area of the closed end, which is the area of the cylindrical cross-section. We determined earlier that the area of the circular cross-section is **4π** square inches. The total surface area is:

40π + 4π

= **44π = 138.16** square inches, based on **π = 3.14**.

If **both** ends of the cylinder are closed, we must add the area of the top and the area of the bottom; that is: the area of the sheet plus twice the area of the cylindrical cross-section:

40π + 4π + 4π

= 48π = 150.72 square inches, based on **π = 3.14**.

20.6.1 Review Problems

Use **π = 3.14**.

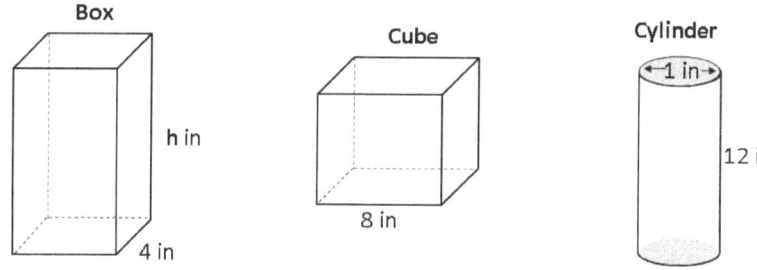

1) The box and the cube in the figure above have the same volume.

   a. What is the height **h** of the box?

   b. What is the difference between the surface area of the box and that of the cube?

2) The cylinder in the figure above is used in a laboratory to hold a solution. It is **12** inches tall and the diameter of its cross-section is **1** inch.

   a. How many cubic inches of solution can it hold?

   b. What is its surface area, if its top is open?

Chapter 21 – Cartesian Coordinate System

21.1 Our Focus

In this chapter, we will discuss how to do rudimentary geometric analysis in the coordinate system. We will focus on points and straight lines. We will cover the slope, equation, and other aspects of straight lines. Finally, we will discuss the graph of a parabola, whose equation is represented by a quadratic equation.

21.2 Cartesian Coordinate System

The Cartesian Coordinate System is named after a French mathematician, René Descartes, who first came up with the idea and developed many of the concepts that we know today. These days, it is simply referred to as the Coordinate System. The coordinate system allows us to do geometric analysis in multi- dimensional space. In our discussion, we will only consider two-dimensional space.

Figure 1: Two-Dimensional Cartesian Coordinate Graph

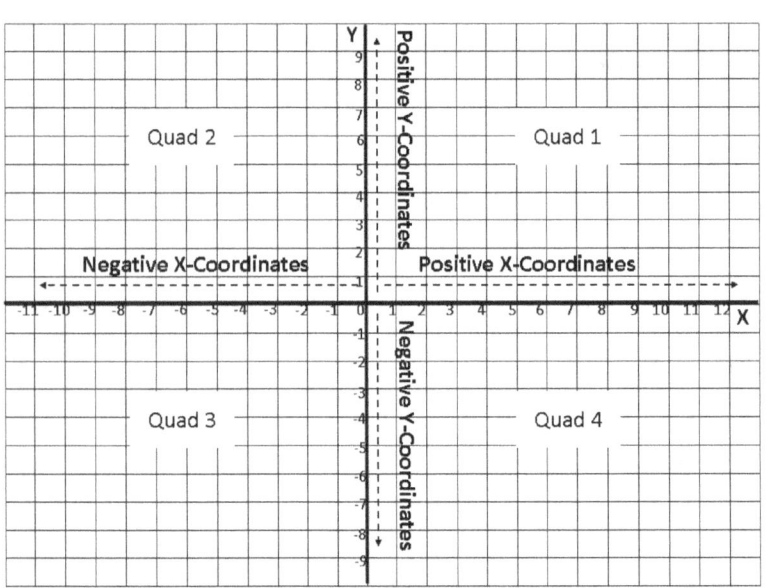

In the coordinate system we represent two-dimensional space as a graph. A graph consists of equally-spaced intersecting horizontal and vertical lines, as shown in *Figure 1*.

Think of two-dimensional space as a plane made up of an infinite number of points. The coordinate system provides a method for plotting and connecting points to make straight lines, curves, and two-dimensional figures of all kinds. With the coordinate system we can do geometric analysis at a depth that would not be possible without it.

In the middle of *Figure 1*, we have two perpendicular number lines. One is horizontal (runs east-west), the other is vertical (runs north-south). The **horizontal** number line is called the **X-axis**. The **vertical** number line is called the **Y-axis**. (You must remember these names, as we will refer to them over and over.)

The point where the X-axis and the Y-axis intersect is referred to as the **Origin**. Equally spaced vertical gridlines mark points on the X-axis. The points are assigned positive integers that increase as we move rightward on the X-axis from the origin, and negative integers that decrease as we move leftward on the X-axis from the origin. Similarly, equally spaced horizontal gridlines mark points on the Y-axis. The points have been assigned positive integers that increase as we move upward on the Y-axis from the origin, and negative integers that decrease as we move downward on the Y-axis from the origin. The origin is always assigned zero.

The two axes (plural for axis) divide the grid into four **quads** as shown in the figure. We count the quads going counter clockwise. Quad One is bounded by the positive X-axis and the positive Y-axis. Quad Two is bounded by the negative X-axis and the positive Y-axis. Quad Three is bounded by the negative X-axis and the negative Y-axis. Quad Four is bounded by the positive X-axis and the negative Y-axis. (Take a minute or two to examine the figure and note all the points we have made about the graph.)

21.3 Coordinates of a Point

Figure 2 shows a point on the graph. The point is labeled **P** and has been assigned the pair of numbers (**3,2**). We have put a dot at the point on the graph to make it stand out.

Figure 2: Coordinates of a Point

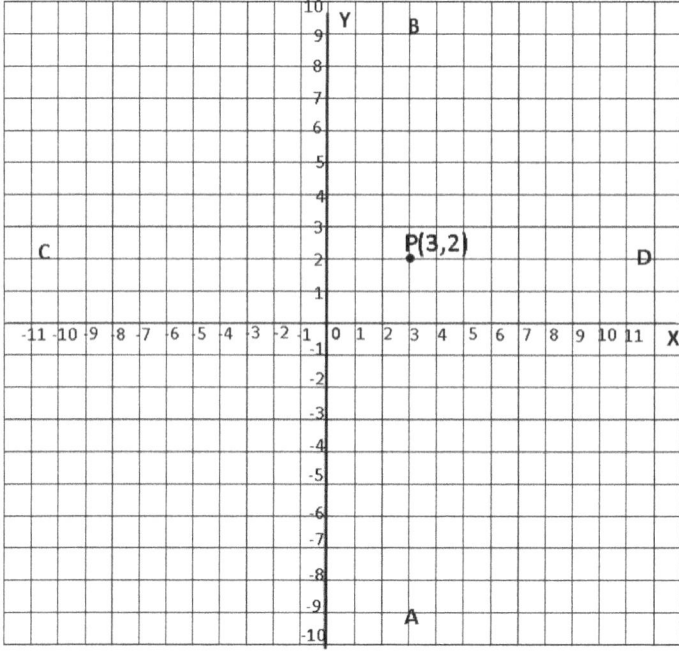

The pair of numbers uniquely identify that point. No other point on the graph will have these identification numbers.

The first number: **3** tells us that the point **P** lines on the vertical gridline that crosses the X-axis at **3**. We have labeled this gridline AB. The second number **2** tells us that the Point **P** also lies on the horizontal gridline that crosses the Y-axis at **2**. We have labeled this gridline CD. In other words, **P** lies at the point where the gridlines AB and CD intersect.

The number pair (**3,2**) is called the **coordinates** of the point **P**. The number (**3**) is the x-coordinate, and (**2**) is the y-coordinate. The coordinates of a point are an **ordered pair**. They follow the order (**x,y**), where 'x' stands for the **x**-coordinate, and 'y' stands for the **y**-coordinate. To identify a point on the grid, we give it a letter label, we write the x-coordinate, a comma, then the y-coordinate. We enclose the coordinate pair in **( )**. Be sure to always follow this rule!

Figure 3: More Coordinates

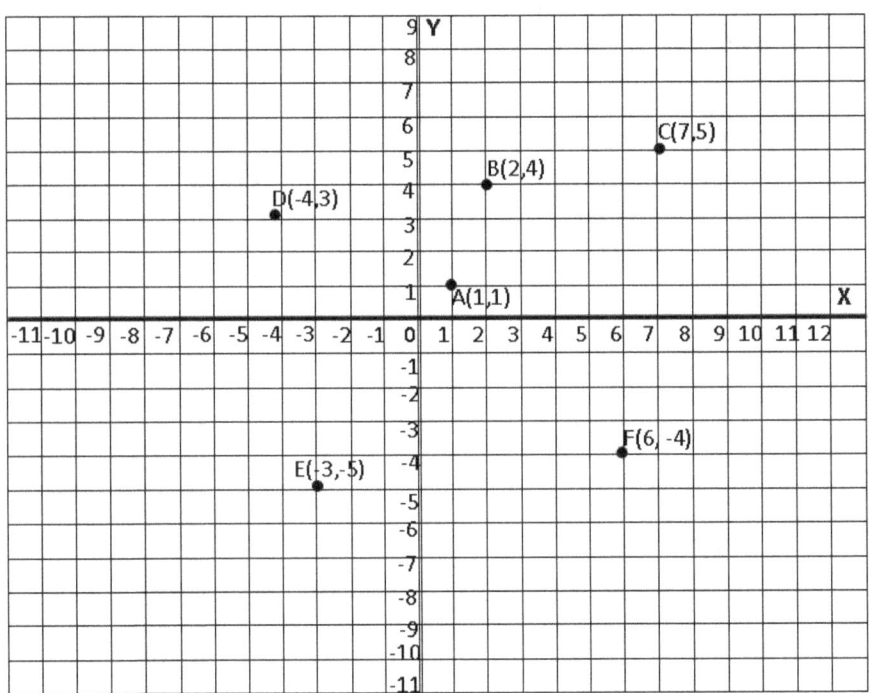

In *Figure 3*, we have plotted several points to further emphasize how we assign an ordered pair of coordinates to a point on a graph.

- The point **A** is at the intersection of x-gridline **1** and y-gridline **1**. Its coordinates are (1,1).
- The point **B** is at the intersection of x-gridline **2** and y-gridline **4**. Its coordinates are (2,4).
- The point **C** is at the intersection of x-gridline **7** and y-gridline **5**. Its coordinates are (7,5).
- The point **D** is at the intersection of x-gridline (**-4**) and y-gridline **3**. Its coordinates are (-4,3).
- The point **E** is at the intersection of x-gridline (**-3**) and y-gridline (**-5**). Its coordinates are (-3,-5).
- The point **F** is at the intersection of x-gridline **6** and y-gridline (**-4**). Its coordinates are (6,-4).

128

Every point that lies on the X-axis itself has y-coordinate zero. Think about it. The X-axis is the horizontal gridline that crosses the Y-axis at the point **0**. Similarly, every point that lies on the Y-axis itself has x-coordinate zero, since the Y-axis is the vertical gridline that crosses the X-axis at the point **0**.

The coordinates of every point in Quad One are both positive. For points in Quad Two the x-coordinate is negative and the y-coordinate is positive. For points in Quad Three, both coordinates are negative. For points in Quad Four the x-coordinate is positive and the y-coordinate is negative. Given the coordinates of a point, we can identify the quad the point lies in.

Again, spend a few minutes to review this discussion and make sure you understand how coordinates are assigned.

21.3.1 Review Problems

i. Make a Copy of Figure 2. Plot the following points on the graph:

B(-5,2), C(3,7), D(-3,-7), E(4,4), F(-4,-4), G(1,-8), H(-8,3),

Q(3,0), V(0,4).

ii. Which Quads do the following points lie in?

iii. Z(2,5), S(-3,-7), O(9,-3), R(7,-2), T(-2,4).

iv. Which of the following points lie on the axes?

Z(0,5), S(-3,0), O(9,0), R(7,-2), T(0,4).

21.4 Equation of a Straight Line

When we connect two points on the graph, we get a straight line. This is shown in *Figure 4*. In the figure, we have connected the points A(1,1) and B(4,4) with a straight line. For each of the two points, the x-coordinate and the y-coordinate have the same value. In fact, the straight line CD, of which AB is a short segment, connects those points on the graph whose x- and y-coordinates have the same value. (Take a close look and confirm this for yourself.) For this reason, the equation of the line is:

y = x

The equation says that for every point that lies on the line CD the y-coordinate equals the x-coordinate. When the x-coordinate is zero, the y-coordinate is also zero. This tells that the line goes through the origin. Remember that the origin is the point where the X-axis and the Y-axis cross, and is the zero point.

Figure 4: Equation of a Straight Line

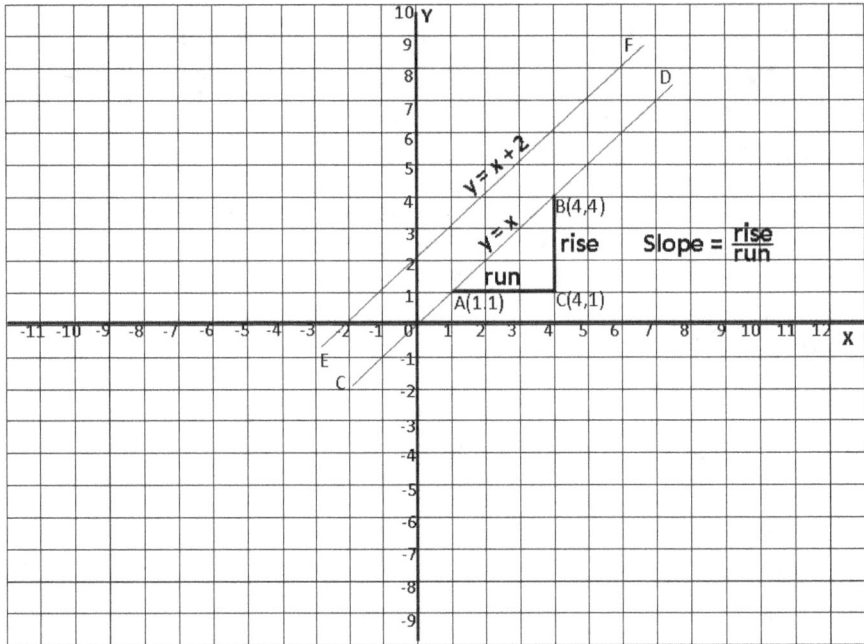

The second line in *Figure 4* has the equation:

$$y = x + 2$$

The equation says: Pick any x-coordinate: **x**, add 2 to it, and you have the y-coordinate: **y**. Every point whose coordinates satisfy this equation lies on the line: **y = x + 2**. Examine the line to verify this for yourself. You will see that point (1,3) lies on it. So does the point (2,4), and (3,5) and so on. For all these points, the y-coordinate is **2** more than the x-coordinate.

In the equation of a straight line, the variable **x** is referred to as the independent variable, and the variable **y** is the dependent variable. All this means is that the value of **y** is derived based on the value we assign to **x**.

In the equation: **y = x + 2**, when x = 0, y = 2. Therefore, the line crosses the Y-axis at the point (**0,2**). Remember that all along the y-axis, the x-coordinate is zero. You can see this from the figure. Another way to say this is that the point (**0,2**) is the **y-intercept** of the line **y = x + 2**.

The line **y = x + 2** crosses the X-axis at (**-2,0**), which is its **x-intercept**.

In the first equation: **y = x**, when **x = 0**, y = 0. This line crosses the Y-axis and the X-axis at the same point: (**0,0**), the origin. The origin is therefore both the y-intercept and the x-intercept of the line: **y = x**.

Check all this out by carefully studying *Figure 4*.

21.4.1 Review Problems

Find the coordinates of two points that lie on each of the following lines:

  a)   y = 2x − 5

  b)   y = 3x + 1

  c)   y = $^1/_2$x + 4

  d)   y = 2x

Determine which line in a) - b) above each point pair lies on:

  i.   C(1,4) and D(-1,-2)

  ii.   A(1,-3) and B(3,1)

  iii.   G(1,2) and H(5,10)

  iv.   E(4,6) and F(8,8)

21.5 Slope of a Line

Let us go back to *Figure 4* and to the line **y** = **x**. In order to calculate the slope of a line we need two points on the line. For line **y** = **x**, we have chosen the two points A(1,1) and B(4,4). The point C(4,1) is where the perpendicular line segments AC and BC meet. Because AC and BC are perpendicular, the triangle ABC is a right triangle. BC is called the **rise** of line segment AB, and AC is called its **run**. The slope of the line is the **rise** divided by the **run**.

We use points B and C to calculate the **rise**. We determine which one of the two points is closer to the X-axis. It is point C. We subtract the Y-coordinate of point C(4,1) from the Y-coordinate of point B(4,4). We get **3** for **rise**.

We use points A and C to calculate the **run**. We determine which one of the two points is closer to the Y-axis. It is point A. We subtract the X-coordinate of point A(1,1) from the X-coordinate of point C(4,1). We get **3** for **run**.

The slope of line **y** = **x** is:

$$^{rise}/_{run} = \, ^3/_3 = 1.$$

We could have chosen any other two points on the line to calculate the slope. In particular, we could have chosen the origin as one point and the point A(1,1) as the other. In this case the **rise** would have simply been the y-coordinate of A, and the **run** would have simply been the x-coordinate of A. Any time we calculate slope using the origin and another point, the rise is the other point's y-coordinate, and the run is the point's x-coordinate. (Do you see why?)

Back to *Figure 4*. You can tell by inspection that the line **y = x + 2** is parallel with the line **y = x**. You can easily verify this by calculating its slope, just like we did for line **y = x**. The slope for the two lines will come out equal. Two lines that have the same slope are **parallel**. Parallel lines belong to a **family** of lines that have the same slope. For each slope, there is an **infinite** number of parallel lines in the family.

21.5.1 Review Problems

Plot the following lines on a graph. Calculate the slope of each line using **rise** and **run**.

   i. $y = 2x - 5$

   ii. $y = 3x + 1$

   iii. $y = \frac{1}{2}x + 4$

   iv. $y = 3x$

   v. $2y = x +$

   vi. From the slopes you calculate, identify the lines that are parallel.

21.6 General Equation of a Straight Line

The following is the general equation of a straight line:

$$y = mx + c$$

The x-term has coefficient **m**, and **c** is a numeric constant. In this form, the coefficient of **y** is **1**. (Remember: **1 × y = y**.) We map corresponding values from the equation of a given straight line to **m** and **c**.

In the form *where the coefficient of y is 1*, the x-coefficient **m** is the slope of the line. The y-intercept of the line is **c** because when **x = 0, y = c**.

Back to *Figure 4*, and the lines: **y = x** and **y = x + 2**.

In the first equation: **y = x,** the value **m = 1,** and **c = 0.** This is consistent with the value **1** that we previously calculated for the slope of the line. This also confirms that the y-intercept of the line is the origin.

In the second equation: **y = x + 2**, the value of **m = 1**, and the **c = 2.** As we indicated above, lines that are parallel have the same slope. So, this confirms that the two lines are parallel. Also, it confirms that the y-intercept of the line **y = x + 2** is the point (0,2).

Sometimes we encounter the equation of a straight line in which the coefficient of the variable **y** is not **1**. Here is an example:

$$3y = 4x + 12$$

In this example, **4** is <u>not</u> the slope, and **12** is not the y-intercept because the equation does not follow the general form. The good news is that it is easy to convert the equation to the general form. In this case, we simply divide each side of the equation by **3**. The equation changes to:

$$y = {^4/_3}x + 4$$

Now, the equation follows the general form. The x-coefficient: ${^4/_3}$ is the slope, and the point (0,4) is the y-intercept.

(As an exercise, you can plot both forms of the equation to verify that they are equations of the same line, and calculate the slope to confirm that it equals ${^4/_3}$.)

Let us move on to **Figure 5**, where we show another straight line CD. Its equation is:

$$y = 2x + 4$$

The equation follows the general form; **m** = 2 and **c** = 4. The line's slope is **2** and its y-intercept is the point (**0,4**). We have also picked two points G(0,4) and E(2,8) to calculate the slope.

Following the rules for calculating **rise** and **run**, we subtract the y-coordinate of the point F(2,**4**) from the y-coordinate of the point E(2,**8**) and get **4** for the **rise**. And we subtract the x-coordinate of the G(0,4) from the x-coordinate of the point F(**2**,4) and get **2** for the **run**. We divide **4** by **2** and get **2** for the slope. This matches the **m** value in the equation. The graph shows that G(0,4) is the y-intercept, and this also matches the **c** value in the equation.

Figure 5: More on the Slope of a Line

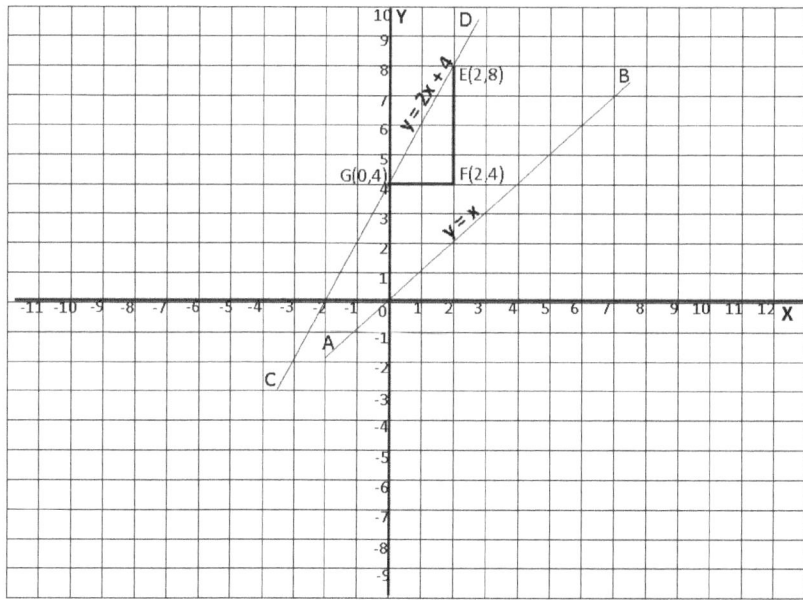

We can tell from the figure that when two lines lie in Quad One, the line with the larger slope is steeper that the line with the smaller slope. We measure steepness relative to the positive X-axis. The steepest a line in Quad One can be is when it points exactly north-south. In that position, its **run** equals zero. When we divide the **rise** of such a line by **run** (zero), the result is infinite or undefined.

When a line lies exactly north south, it is parallel with the Y-axis, and it is perpendicular to the X-axis. This tells us that the slope of the **right angle** is **infinite**.

21.6.1 Review Problems

Find the slopes and y-intercepts of the following lines:

   i. $2y = 4x + 8$

   ii. $y = 17x + 11$

   iii. $4y = 3x + 20$

   iv. $y = {}^5/_2x - 12$

21.7 Finding the Equation of a Straight Line

An infinite number of lines can go through any one point on a graph. ***Figure 6*** dramatizes this. It shows four lines going through the point C(4,4). However, two points A(-6,5) and B(-4,2) define a unique straight line, as shown in the same figure.

Figure 6: Lines Through a Point

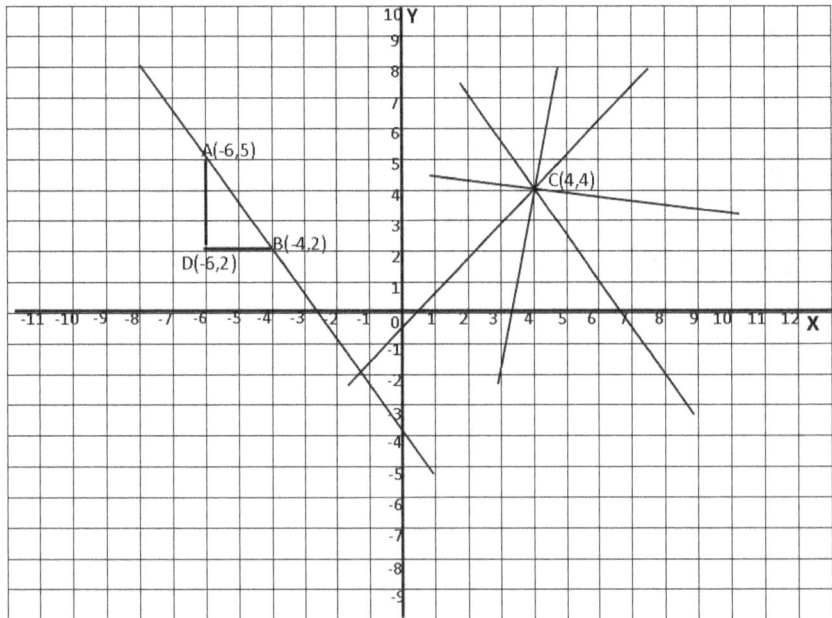

Quite often we have to determine the equation of a straight line. We need the coordinates of two points on the line, or the coordinates of one point on the line in addition to the line's slope.

21.7.1 Equation of Line from a Point and Slope

Let us first determine the equation of a straight line when we know one point on the line and the line's slope.

Example 1: Equation of the line with the point A(3,2) and slope **5**.

We start with the general equation of a straight line:

$y = mx + c$.

We know the slope of the line in *Example 1*; we substitute it for **m** in the general equation. We get:

$y = 5x + c$

Since the point A(3,2) lies on the line, we substitute **3** for **x**, and **2** for **y**. We get:

$2 = 15 + c$

This gives:

$-13 = c$

Therefore, the equation of the straight line is:

$y = 5x - 13$

We can easily verify that the point A(3,2) lies on this line.

21.7.2 Equation of a Line from Two Points

Now, let us determine the equation of a straight line when we know two points.

Example 2: Equation of the line that goes through the points A(-6,5)

and B(-4,2).

This is the line in Quad Two of *Figure 6*.

Again, we start with the general equation of a straight line:

$y = mx + c$.

Since the line goes through A(-6,5), we replace the variables **x** and **y** with the coordinates of A. We get:

Equation 1: **5 = -6m + c**

We do the same with the coordinates of the point B(-4,2). We get:

Equation 2: **2 = -4m + c**

We have two linear simultaneous equations which we solve for **m** and **c**, using the techniques we learned previously. Let us go right through.

We subtract the right side of **Equation 2** from the right side of **Equation 1** and get:

$-6m + c - (-4m + c)$

$= -6m + c + 4m - c$

$= -2m$

We subtract the left side of **Equation 2** from the left side of **Equation 1** and get:

$5 - 2 = 3$

Therefore:

$3 = -2m$

$-^3/_2 = m$

We substitute: $-^3/_2$ for: **m** in Equation 2 and solve for **c**:

$2 = -4 \times -^3/_2 + c$

$2 = 6 + c$

$-4 = c$

We substitute the values **m** $= -^3/_2$ and **c** $= -4$ into the general equation of a straight and get;

$y = -^3/_2 x + (-4)$, which we simplify and get:

$y = -^3/_2 x - 4$

This is the equation of the line. (We can write it as: **2y = -3x − 8**)

From the figure, we can tell immediately that the y-intercept is the point (0,-4). This is consistent with the equation of the line.

Now, let us calculate the slope using **rise** and **run** from the graph.

The point D(-6,2) is closer to the X-axis than the point A(-6,5). We subtract the y-coordinate of D from the y-coordinate of A and the **rise** is:

$5 - 2 = 3.$

The point B(-4,2) is closer to the Y-axis than the point D(-6,2). We subtract the x-coordinate of B from the x-coordinate of D and the **run** is:

$-6 - (-4) = -6 + 4 = -2$

Therefore, the slope of the line is: **3** $\div$ **(-2)** $= -^3/_2.$

The values match the solutions for **m** and **c** in the simultaneous equations.

Note: It is not an accident that the slope of the line is negative. We will discuss angles and the signs of slopes next.

21.7.3 Review Problems

Find the equations of the lines that each of the following pairs of points lie on:

    i.   A(1,4) and B(-1,-2)

    ii.   C(1,-3) and D(3,1)

    iii.   E(1,2) and F(5,10)

    iv.   G(4,6) and H(8,8)

    v.   I(1,6) and J(2,13)

21.8 Angle and Slope

In *Figure 7*, the line OB sweeps counter-clockwise around the circle through 360°. It starts the sweep when it is coincident with the X-axis. When it completes the sweep through 360°, it again becomes coincident with the X-axis.

Figure 7: Angle and Slope

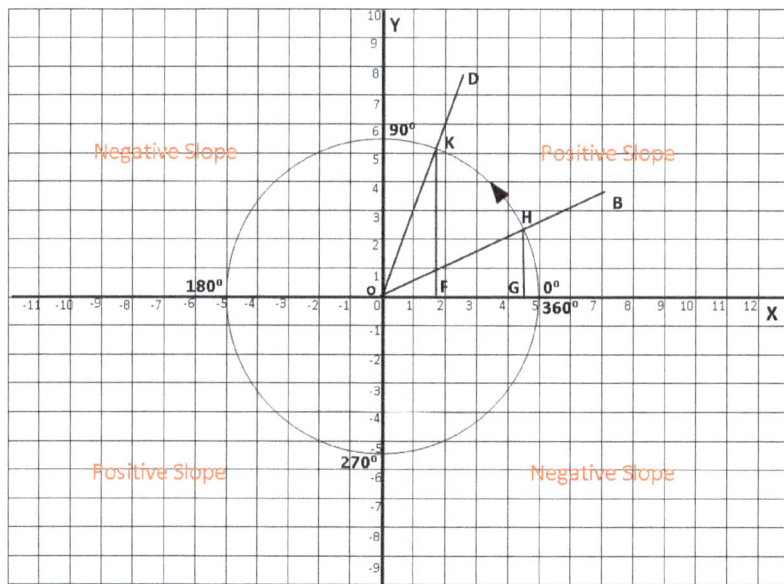

At the start, when the OB is coincident with the X-axis, it makes an angle **0°** with the X-axis. At that point, its **rise** is also zero because it is literally lying on top of the X-axis.

The line continues its sweep and is now in the position OB. It now makes angle ∠HOF with the X-axis and its **rise** has increased from zero to some positive value. It continues its sweep and gets into the position

OD. The angle it now makes with the X-axis is ∠KOF, which is much larger than ∠HOF. The **rise** has increased even more and the **run** has become much less than in the line's previous position OB. This means that its slope has increased even more rapidly.

When the line coincides with the Y-axis, it has swept through 90°. Its **run** in now zero (because it is literally lying on the Y-axis at this point), and its slope is ∞. (When we divide any number by zero the result is infinite: ∞.) As the line continues its sweep from 90° to 180°, its slope goes from ∞ to zero again. In between 90° and 180°, the slope is negative. For example, its slope at 135° is **-1**. (Remember that we measure angle relative to the positive X-axis.)

When the line gets to 180°, it becomes coincident with the X-axis a second time and its slope drops to zero. It continues the sweep from 180° to 270°. Between 180° and 270° its slope is positive and take on the same values as between 0° and 90°. For example, at 225°,

its slope is **1**, same value as at 45°.

At 270°, the line is coincident with the Y-axis a second time, and its slope is again infinite. It continues its sweep from 270° to 360°. Between 270° and 360°, its slope is negative and takes on the same values as between 90° and 180°. For example, at 315°, its slope is **-1**, same value as at 135°.

When the line reaches 360°, it has completed the revolution and is back to being coincident with the positive X-axis. Thus its slope at 360° is zero, same value as at 0°.

Here are key points from the preceding discussion:

- We determine the angle of a line relative to the positive X-axis.

- The slope of a line alone is not enough to know the angle the line makes with the X-axis because in a 360° sweep, two different angles produce the same slope. This is one reason why slope alone is not enough to determine the equation of a line.

21.9 An Application of Linear Equations

Two enterprising teenagers, Jon and Mark, make spending money by planting shrubs for their neighborhood homes. Jon charges $10 for the service call and $5 per shrub planted. Mark charges a flat $7.50 per shrub planted, with no service call fee.

Mrs. Sousa wants to plant shrubs along the front of her property to make a natural fence. Should she hire Jon or Mark to do her planting? Let us analyze the problem.

Mark's fee schedule is quite simple: $7.50 × (Number of shrubs planted). The equation for it is:

$$y = 7.5x$$

where **x** is the number of shrubs Mark plants, and **y** is the fee he charges. There is zero fee when Mark plants zero shrubs because he does not charge for a service call.

Jon, on the other hand, charges $10 when he plants zero shrubs. This is his service call fee. Then he charges $5 for each shrub he plants. The equation for Jon's fee schedule is:

$$y = 5x + 10$$

where **x** is the number of shrubs Jon plants, and **y** is the fee he charges.

The lines in **Figure 8** represent the fee schedules of the two boys. The lines are labeled by the boys' names. In the figure, the X-axis represents the number of shrubs planted, and the Y-axis represents the fee paid.

Figure 8: Fee Schedules for Jon and Mark

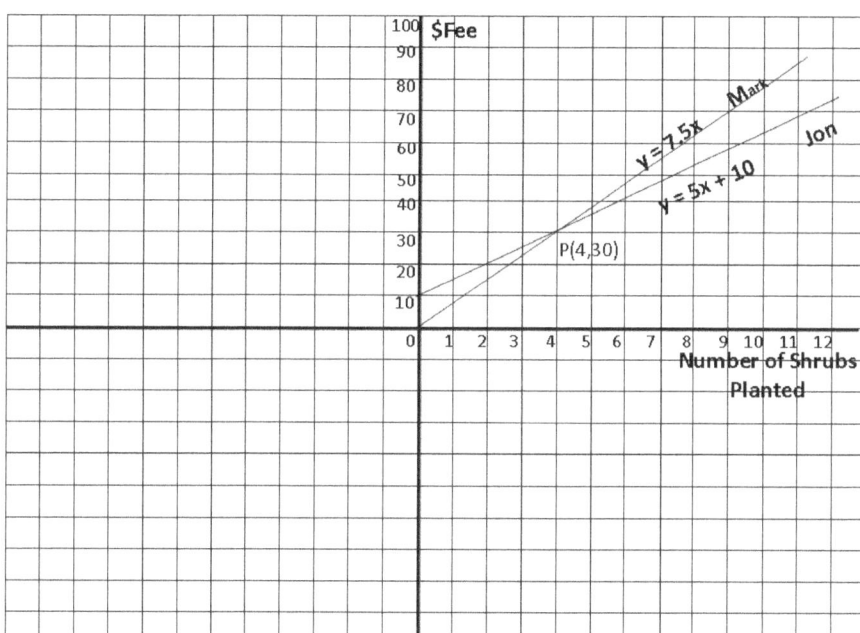

We see from the graph that the two lines intersect at the point P(4,30). This tells us that the fee for planting 4 shrubs is $30, and it is the same for both boys. If Mrs. Sousa intends to plant 4 shrubs, it does not matter which boy she uses. Below 4 shrubs, Mrs. Sousa will pay less by going with Mark. You can see that for this range, Mark's line lies below Jon's. For example, at 2 shrubs, Mark's charges are $15; Jon's charges $20. (Note that the Y-axis scale is in increments of $10.) However, from 5 shrubs and up, Mrs. Sousa is better off contracting Jon. For the range of 5 and up, Jon's line lies below Marks.

We can also determine the point P(4,30), by solving the simultaneous equations:

Equation 1: $y = 7.5x$

Equation 2: $y = 5x + 10$

The question of which boy to choose for the work is a linear problem. It means that the facts conform to the equation of a straight line, as we can see from **Figure 8**. We can use graphs of straight lines to analyze and solve linear problems.

Using a graph like **Figure 8**, Mrs. Sousa will know instantly how the fee schedule plays out for the two boys (y-value) for any number of shrubs she needs planted (x-value).

21.9.1 Review Problems

i.   Solve the simultaneous equations for Mark's and Jon's fee schedule to confirm the solution **x** = 4, **y** = 30.

ii.   Mr. Lee is taking his family on a 10-day vacation in Florida and will need a rental car while in Florida. The rental car agency, Blue Inc., charges a flat $75 per day. The other agency, Green, In., charges $60 per day, plus 0.10 per mile for the same class of car. Which agency should Mr. Lee rent from? His decision is based on the number of miles he expects to drive. Use equations of a straight line to analyze the decision facing Mr. Lee.

21.10   Parabola - Plot of a Quadratic Function

Earlier, we discussed quadratic equations and their solutions. We looked at the general form of the quadratic equation, which we repeat in **Equation 1**.

$$\textit{Equation 1}: \quad ax^2 + bx + c = 0$$

In this form, we are only interested in the two values of **x** that reduce the left side of the equation to zero.

Now, let us look at the following equation:

$$\textit{Equation 2}: \quad y = ax^2 + bx + c$$

In **Equation 2**, we are no longer just interested in the zero value. Every value of **x** we pick will produce a corresponding value for **y**. We say that the variable **y is a function of x**; that is, **y** depends on **x** for its values. We refer to **x** as the **independent** variable, and **y** as the **dependent** variable. (We encountered this concept when we discussed the equation of a straight line.) The following is the mathematical shorthand for "**y is a function of x**": **y = f(x)**.

In **Equation 2**,

$$f(x) = ax^2 + bx + c$$

Think of **f(x)** as a calculator with the polynomial representing the routine that **f(x)** uses to calculate values for **y**.

When we set **y** in **Equation 2** equal to zero, it takes us back to **Equation 1**, where we ask the question: "What values of **x** make **y** zero?"

When we plot a quadratic function of the **x-y** graph, we get a curve of the shape shown in **Figure 9**. The curve is called a **parabola**. It can be open at the top, like the one shown, or open at the bottom (ie., turned upside down.) It depends on equation of the polynomial used to plot it. In the figure, we have omitted the letter names of the identified points simply to make the graph less busy.

Figure 9: X-Y Plot of a Quadratic Function

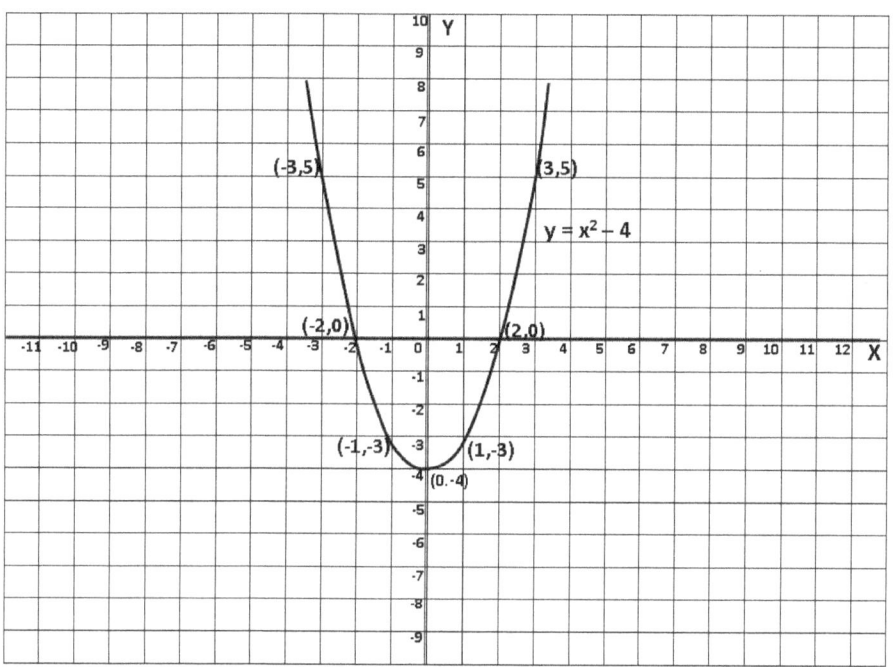

The parabola in the **Figure 9** has the equation:

$$y = x^2 - 4$$

Let us look at a few values of **x** and their corresponding **y** values.

x	0	1	2	3	4	5
x		-1	-2	-3	-4	-5
y = f(x)	-4	-3	0	5	12	21

The table shows **x** values from **-5** to **5**, and the corresponding values of **y**. We notice the following:

> **x = -1** and **x = 1**, generate the same value: **y = -3**,

> **x = -2** and **x = 2**, generate the same value **y = 0**, and so on.

This is what creates the parabola's symmetric shape, and it's the direct result of the x^2 term in the equation. Squaring a positive number and its negative gives the same positive result.

We see from the figure that the curve of **y = x^2 - 4** crosses the x-axis at the points (-2,0) and (2,0). In other words, when **x = -2** or **2, y = 0**.

When **y = 0**, the equation becomes:

$$x^2 - 4 = 0$$

Using the factoring technique we learned previously, we arrive at:

$$(x + 2)(x - 2) = 0$$

The solution is: **x = -2, x = 2**.

(Another approach: since $x^2 - 4 = 0$, $x = \sqrt{4} = \pm 2$)

We can see that the solution corresponds to the same points at which the parabola crosses the x-axis in **Figure 9**. And we already know that at those points the y-coordinate equals zero.

When we set **y = 0** in any function **y = f(x)** and solve for **x** in **f(x) = 0**, we are determining the point(s) where the function's plot intersects the x-axis. If the plot does not intersect the x-axis, there will be no solution for **f(x) = 0**.

The equation:

$$y = x^2 + 4$$

looks similar to the equation of the parabola in **Figure 9**. However, in this case, when **x = 0, y = 4**. The base of the parabola sits on the point (0,4), which is above the x-axis. This parabola does not intersect the x-axis. When we try to solve the equation for **y = 0**, we get:

$$x^2 = -4; x = \sqrt{-4}$$

The square root of a negative is not a real number, so, for our purpose, a solution does **not** exist. (Plot **y** $= x^2 + 4$ on a x-y graph to see what it looks like.)

If **f(x)** represents a straight line, there will be one such point, if the line does cross the x-axis. Not every straight line does. For example, the straight line: **y = 2** does not cross the x-axis.

If **f(x)** represents a quadratic function which crosses the x-axis, there may be two such points. In that case, the solution of the equation will have two values. We have seen this from our examples and review problems.

If **f(x)** represents a third order polynomial, there may be three such points. In that case, the solution to the equation will have three values.

21.10.1 Review Problems

Plot the following functions on a **x-y** graph. Determine the x-intercept(s) for each function. Then solve each equation when **f(x) = 0**, using the techniques we have learned. Verify that the solution(s) matches the x-intercept(s) in each case.

i. $y = f(x) = x + 6$

ii. $y = f(x) = \frac{1}{2}x - 4$

iii. $y = f(x) = x^2 - 2x - 3$

iv. $y = f(x) = x^2 - 1$

Chapter 22 - Graphs

22.1 Our Focus

Graphs give data a pictorial perspective. They help us to quickly get a sense of how the elements of the data under review relate to each other. They save some of the tedious work of analyzing data in order to make sense out of it. Additionally, because graphs are pictures, they add an emotional element to how we perceive the information being presented. In this chapter, we will discuss three of the most popular types of graphs: **Histogram**, **Pie Chart**, and **Line Chart**.

22.2 The Histogram

The **Histogram** is used to compare data or results that are mutually independent. The important word here is "independent". For example, businesses often use histograms to compare, present, or report on the performance of different divisions of the company. The superintendent of schools may use a histogram to compare the standard test performance of various senior classes of the high schools under his watch. In both cases, the performance of one group is independent of the others. How many gadgets a company's division in California sells is independent of how many its New York division sells. And, how one senior class in one town performs on a standardized test is independent of how another senior class in another town does.

Figure 1 is an example of a histogram. We will use it for our discussion. The histogram presents the quarterly earnings of a fictitious company called Pools, Inc., which has three divisions: East Division, West Division and North Division. The histogram shows the company's quarterly earnings for the year 2002. Each quarter's report is broken down by division. The legend on the right uses colors to associate each division with its earnings for the quarter: light blue for East Division, dark blue for West Division, and green for North Division.

The quarters of the year 2002 are labeled below the horizontal line at the bottom of the graph. The scale for the earnings is shown on the vertical line on the left of the graph. Each spike sits on the quarter it belongs to and the height of the spike indicates the value it represents. In this example the value is **earnings** in thousands of dollars.

This is how a histogram is typically constructed. The categories to be compared are labeled on the horizontal base and the values associated with the categories are shown on a vertical scale on the left. We use a scale that allows us to show the smallest value as well as the largest value. In *Figure 1*, the scale is in increments of $10 thousand. For example, between 10 and 20, we have 11, 12, 13, ...,19, which are not shown, but are implied. We keep this in mind when we determine amount from the height of a spike.

Different colors or different shades of color are used to distinguish data for one category from the next. We try to be creative when we design of a histogram, but the design should not distort the accuracy of the data being reported.

Figure 1: Year 2002 Quarterly Earnings of Pools, Inc.

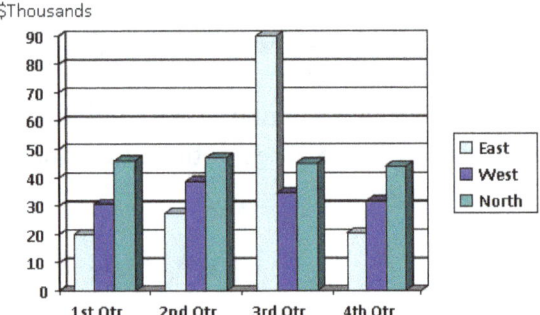

Some things immediately stand out in the histogram:

- East Division was the worst performer for three of the four quarters, but in the 3rd quarter, it rocketed up in earnings.
- Without even adding the numbers, we can tell that the best earnings performance for the company as a whole occurred in the 3rd quarter.
- The earnings performance for the company as a whole is roughly the same during the 1st and 4th quarters and is the worst of 2002.
- If we assume that Pools, Inc., is based in the USA, we can conjecture that perhaps its business is seasonal, with demand dropping off during the fall and winter months. This would account for the earnings performance drop-off during the 1st and 4th quarters.

See what other observations you yourself can make.

A good graph allows us to gleam the key information being presented at a glance. For a busy executive, that is an invaluable time saver. For this reason, histograms are popular with businesses.

If you are the manager of East Division, seeing your division's earnings for the 3rd quarter tower above the earnings of the other two divisions of the company will make you proud. You can see how powerful a tool a graph can be.

22.2.1 Review Problems

Use *Figure 1* to answer the following questions:

    I. Approximately how much was the single highest quarterly revenue reported by a division?

II.   Approximately how much was the lowest quarterly revenue reported by a division in any quarter?

III.   Approximately how much was the company's highest combined quarterly revenue?

22.3 The Pie Chart

The Pie Chart is a circular graph that resembles a pie, as shown in *Figure 2*. That is the reason for its name.

Figure 2: Pie Chart - Mary's Monthly Expenses

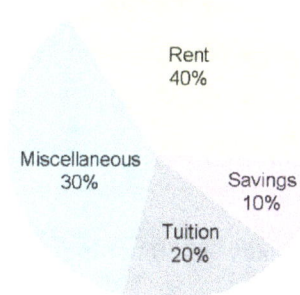

The information in a Pie Chart is organized in slices, just as a pie is cut in slices. Typically, different colors, or shades of color, are used to differentiate the slices, as shown in *Figure 2*.

We use a **pie chart** to present information whose parts are related to each other and have mutual dependence. Parts of information are inter-related and mutually dependent if tweaking one part has an effect on another part.

Figure 2 is a pie chart showing how Mary uses her monthly income. Mary is a fictitious young woman living and working in New York City. To improve herself, Mary goes to school at night. She pays her own tuition. Her other expenses include rent, her monthly savings, and what she calls miscellaneous expenses. The latter covers food, transportation, entertainment, clothes, and so on. From the pie chart, the following is the allocation of Mary's income, as a percentage of the total:

Rent:	40%
Miscellaneous:	30%
Tuition:	20%
Savings:	10%
-------------------	-------
Total	**100%**

Obviously, if Mary's monthly income stays the same and any of her allocations increases or decreases, it will impact one or more of the other categories. For example, if her tuition expense goes up, she may have to reduce her savings. If her rent increases, she may have to reduce her allocation for miscellaneous expenses. You can see the inter-relationship and dependence.

The pie chart allows us to instantly tell where Mary's monthly income goes, and how much goes where. The complete pie represents one hundred percent of her expenses. The question of how much of it goes where is answered by the size of each slice, which is proportional to the percentage of the total that is allocated to it. The allocations add up to one hundred percent of the total. That is why a decrease or increase in one category shows up somewhere else to account for one hundred percent of the total.

We can tell from **Figure 2** that Mary's biggest expense is rent, which accounts for 40% of her expenses. The size of the rent slice indicates as much. Her monthly savings represent 10% of the total. Again, the size of the savings slice reflects this.

We can see that just like the histogram, the pie chart also captures and presents key information in a powerful way.

22.3.1 Review Problems

Use the Pie Chart in Figure 2 to answer the following questions. Assume Mary makes $5050 per month.

I.   How much does she save every month?

II. If her tuition expense increases to 24% of her monthly salary and she takes money from the category Miscellaneous to cover the increased tuition, how much spending money will be left in Miscellaneous?

22.4 The Line Chart

The line chart is used to show a trend; typically, a trend over some time period. If you have ever looked at the financial pages of the Wall Street Journal or the New York Times, you have seen line charts. They are used to show how various investment instruments such as stocks, bonds, etc., have performed over some period of time. This gives investors a form of guidance.

Figure 3: Line Chart - Juan's Assist Count

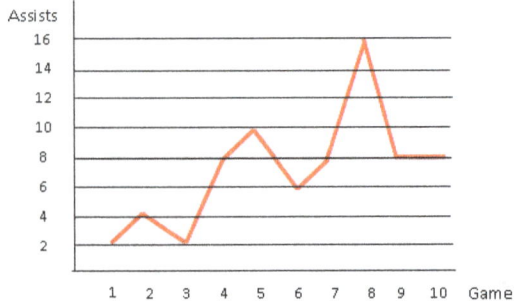

Figure 3 is an example of a line chart. It shows the number of assists the guard, Juan, on a fictitious high school basketball team had per game over a ten game period. An assist is a pass to a teammate that immediately results in a score. It is one measure of a guard's ability and contribution to the team.

The horizontal line shows the game number of each game, from the **1st** game to the **10th**. The vertical line shows the number of assists Juan recorded in each game. For example, if we go to **Game 2** and move vertically up to the line, we find that Juan had **4** assists in that game.

Connecting the points that match assists to games with lines allows us to immediately see the trend in Juan's assist performance. Even though the line zigzags, we can still see that his assist count rose over the 10 games.

Clearly, the rise was not smooth. For example, between games 5 and 6, his assist count dropped by 4. The biggest rise in Juan's assist count for consecutive (back-to-back) games was between game 7 and 8, and the biggest drop in assists between consecutive games was between game 8 and game 9. We can tell all this by simply looking at the shape of the line. Again, this confirms the power of a graph.

We can imagine the coach of the team sitting down with Juan over a line graph like this one discussing the reasons for the variations in Juan's assist performance.

22.4.1 Review Problems

Use *Figure 3* to answer the following questions:

    I.   What was Juan's biggest jump in assists between consecutive games?

    II.   What was Juan's second biggest drop in assists between consecutive games?

    III.   How many times did Juan have a two-assist improvement between consecutive games?

    IV.   Was there an increase in Juan's assist count between from game 9 to game 10?

Chapter 23 – Basic Statistics

23.1 Our Focus

In this chapter, we will discuss a few simple concepts in Statistics. We will focus on some of the terms we commonly encounter, such as Arithmetic Mean, Range, Mode, Median, and so on. We will learn how to calculate the Arithmetic Mean and the Mean Absolute Deviation. We will only skim the surface because to go any deeper will require more advanced mathematics than we have covered. We will also discuss a few simple concepts in Probability.

23.2 What we Learn from Statistics

Statistics is the application of mathematics to analyze data about events in order to uncover characteristics such as trends, inter-dependencies, randomness, bias or skew, cycles, and so on. Thus informed, we are able to make predictions about future events. When you hear on the radio that a particular politician is behind or leading an election campaign by some percentage, a statistician is behind the report.

Climate scientists use statistics to study trends in weather events, such as temperature, heat, rainfall, storm activity, and so on. Epidemiologists use statistics to study patterns, causes and spread of diseases, and use the information to help formulate public policy on health issues. Environmental scientists use statistics to study the impact changes in the environment exert on life on our planet. The application of statistics is pretty much limitless.

The histogram, pie chart, and line chart that we previously discussed are some of the tools statisticians use to report the outcomes of their analysis.

23.3 Some Basic Elements of Statistics

Before we start our discussion, let us set the stage by creating some data.

Imagine that a boy walks home from school every day. School ends at 3:00 pm. The boy's mom sets a timer promptly at 3:00 pm every day to record the boy's travel time from school to home.

Here is the data for the boy's travel time for a three-week period (15 days). The time is in minutes. In statistics, each data point is referred to as a **sample** or **observation**. We have 15 observations in our example.

A proper statistical study of the boy's travel time would require many more samples than 15, perhaps hundreds. In statistics, the more samples we use the more accurately the results reflect reality. This is because a large number of samples allows random variations between samples to be smoothed out.

Travel time observations for boy: {**20, 18, 20, 15, 23, 30, 20, 21, 19, 22, 20, 19, 24, 20, 18**}. Number of observations: **15**.

The observations have some obvious attributes:

- The boy's travel time varies, reflecting the random variations between samples we mentioned above.
- His shortest travel time is 15 minutes.
- His longest travel time is 30 minutes.
- Travel time of 20 minutes has the highest frequency.
- The longest travel time (30 minutes) is twice the shortest travel time (15 minutes).

We do not expect the boy to make it home every day in exactly the same time. There are bound to be delays at school and distractions along the way that cause variations. However, the 15-minute and 30-minute samples stand out.

What could be behind the 15-minute travel time? Perhaps the boy got a ride part of the way. Perhaps he run home because of an imminent storm. In any case, it does not appear to fit the norm. Similarly, the 30-minute sample does not appear to fit the norm. Did the boy start for home then go back to school to retrieve something he had forgotten? Was there an event at school that caused him to have a late start for home?

A statistician would most likely set aside 15-minute and 30-minute travel times as **outliers**. For our purposes, we will include them both.

Now, let us start with some statistical terms.

23.4 Mode

The sample that occurs most frequently in a set is called the **Mode**. As we observed above, travel time of **20** minutes has the highest frequency in our 15 samples. Thus, 20 minutes is the Mode of our set of observations.

If two or more samples occur the most and have the same frequency, the set of observations is said to be **multimodal**.

23.5 Median

Ordered samples: 15, 18, 18, 19, 19, 20, 20, 20, 20, 21, 22, 23, 24, 20, 30.

Above we have sorted our samples from the smallest to the largest. The middle position of the sorted list is position 8. The sample at that position is 20 minutes. In statistics, after we sort a set of observations from smallest to largest, the number in the middle position is called the **Median**. In our case, 20 is the median. (It is a coincidence that 20 is also the Mode.)

We have an odd number of observations in our list: 15. As a result, we have an exact middle position. When the number of observations is even, we end up with two observations in the middle positions. To get the median, we sum the two middle observations of the sorted list and divide the result by 2. In such a case, the median may not match any of the actual observations.

The median gives us a sense of the "middle of the pack" of the set of samples. News outlets often report the **median income** is a given geographic area of the country. The report is meant to help people in the covered area understand by how much they lag or lead the middle of the pack of income earners in their area.

23.6 Arithmetic Mean

If we add the data samples and divide by the count of samples, we get the **Arithmetic Mean**.

Arithmetic Mean = Sum of the observations ÷ number of observations.

The arithmetic mean is commonly referred as the **Average**. The average for our sample is:

(20+18+20+15+23+30+20+21+19+22+20+19+24+20+18) = 309

309 ÷ 15 = **20.6** minutes.

On the average, the boy takes **20.6** minutes (or 20 minutes and 36 seconds) to get home.

What does the arithmetic mean tell us? If we had many more samples, the arithmetic mean would be a good predictor of the boy's travel time. In a small set of observations, like ours, the variations in the samples have a big impact on the arithmetic mean. In a much larger sample, the variations get evened out when we calculate the arithmetic mean.

23.7 Range

The **Range** is the difference between the smallest sample and the largest sample. The range for the boy's travel time observations is: **30 – 15** = **15** minutes.

If one of the boy's friends stops by the boy's house at 3:00 pm and asks his mom how soon the boy will be home, the mom may say **15** to **30** minutes. She will use the **range** to give a safe response. However, when she has half a schoolyear's worth of observations, she may notice that the average constitutes a more precise predictor of her son's travel time.

23.8 Mean Absolute Deviation

To calculate the **Mean Absolute Deviation**, we first subtract each sample from the average. Since some samples are larger than the average, some of the differences will be negative. We sum the **absolute values** of the differences, then divide the sum by the number of observations.

Table 1: Absolute Mean Deviation

Observation Number	Observation Value	Difference from the Average	Absolute Value of Difference
1	20	20.6 – 20 = 0.6	|0.6| = 0.6
2	18	20.6 – 18 = 2.6	|2.6| = 2.6
3	20	20.6 – 20 = 0.6	|0.6| = 0.6
4	15	20.6 – 15 = 5.6	|5.6| = 5.6
5	23	20.6 – 23 = -2.4	|-2.4| = 2.4
6	30	20.6 – 30 = -9.4	|-9.4| = 9.4

7	20	20.6 − 20 = 0.6	|0.6| = 0.6
8	21	20.6 − 21 = -04	|-0.4| = 0.4
9	19	20.6 − 19 = 1.6	|1.6| = 1.6
10	22	20.6 − 22 = -1.4	|-2.4| = 1.4
11	20	20.6 − 20 = 0.6	|0.6| = 0.6
12	19	20.6 − 19 = 1.6	|1.6| = 1.6
13	24	20.6 − 24 = -3.4	|-3.4| = 3.4
14	20	20.6 − 20 = 0.6	|0.6| = 0.6
15	18	20.6 − 18 = 2.6	|2.6| = 2.6
			Sum = 34
			34 ÷ 15 = 2.27

The table shows the calculation of the Mean Absolute Deviation for our set of observations. Its value is: **2.27** minutes.

If we had a much larger number of observations, we would see many more samples that closely track the boy's true travel time. We would see many small absolute differences; they would contribute to a comparatively smaller sum. Also, the sum would be spread over a much larger number of observations. The resulting Mean Absolute Deviation would likely be much less than 2.27 minutes.

What does the Mean Absolute Deviation tell us? It tells us about how precise or imprecise the arithmetic mean is as a predictor. In a small sample, like ours, the mean absolute deviation tends to be a relatively large percentage of the average. In our case it is 11%. The lower this percentage is, the closer to reality the observations and the average are. Clearly, saying that something will happen in 20.6 minutes, give or take 2.7 minutes, is not as precise as 20.6 minutes give or take 0.5 minutes. In statistics we want to be as precise as we possibly can. The larger the number of observations the more precision we achieve.

23.9 Basics of Probability

Probability comes into play when for a given event there are many possible independent outcomes. If each outcome has an equal chance of occurring, the outcomes are said to be **random**. For example, when we toss a fair coin, two outcomes are possible: heads or tails. Each outcome is independent of the other and has an equal chance of occurring.

Since a coin toss has two possible independent equal-chance outcomes, each outcome has half a chance. In statistics, we say that the **Probability** of heads is $^1/_2$ and the **Probability** of tails is also $^1/_2$.

For a set of independent equal-chance outcomes, we define probability as follows:

Probability of an Outcome = Expected Outcome ÷ Total outcomes.

Expected Outcome may be one individual outcome or group of individual outcomes.

The **sum** of the individual probabilities of all the possible outcomes always equals **1**. A probability of **1** indicates 100% certainty that the event will occur. If you hear the weatherman report 100% probability of rain, it is probably already raining.

Let us look at more examples. When we roll a dice, we expect 1, 2, 3, 4, 5, or 6 to show. We have a set of six outcomes. On any roll, the expected outcome is one of six possibilities. The probability that **2** will show is: $^1/_6$. Each of the other numbers has the same probability of showing.

What is the probability that on rolling of a dice any one out of the three numbers: 1, 2, 5 will show? This time, we have 3 out of 6 chances to get a desired outcome. The probability is $^3/_6 = ^1/_2$. It is the **sum** of the individual probabilities of the expected outcomes in the group.

There are 52 cards in a deck. If the card has been thoroughly shuffled, the probability of drawing any one particular card from the deck is $^1/_{52}$.

Probability of drawing the King of Diamonds: $^1/_{52}$.

However, if we simply want to draw any diamond, we have 13 chances out of 52, since there are 13 diamonds in a full deck. Each card in the diamond suit has $^1/_{52}$ chance of being drawn and any one of them will do. Therefore, the probability of drawing a diamond out of the deck of 52 cards is: $^{13}/_{52}$.

23.10 Multiplication of Probability

What is the probability of tossing a coin twice and getting ***heads-heads***? When outcomes are independent, we determine the probability that multiple outcomes will occur in a **predetermined sequence** by **multiplying** the probabilities of the outcomes in the sequence. In a coin toss:

Probability of Heads-Heads: $^1/_2 \times ^1/_2 = ^1/_4$

Probability of Heads-Tails: $^1/_2 \times ^1/_2 = ^1/_4$

Probability of Tails-Heads: $^1/_2 \times ^1/_2 = ^1/_4$

Probability of Heads-Tails-Tails: $^1/_2 \times ^1/_2 \times ^1/_2 = ^1/_8$

If we make a bet to draw any diamond (**13** chances out of **52**) from a well shuffled deck of cards, put it back into the deck, reshuffle the deck and draw the Jack of hearts (**1** chance out of **52**), our probability of success is:

$^{13}/_{52} \times ^1/_{52} = ^{13}/_{2704} = 0.005$ (half of 1%)

If we make a bet to draw a diamond (**13** chances out of **52**) from a well shuffled deck of cards, put it back into the deck, reshuffle the deck and draw any spade (**13** chance out of **52**), our probability of success is:

$^{13}/_{52} \times ^{13}/_{52} = ^{169}/_{2704} = 0.0625$ ($6^1/_4$%)

23.11 Addition of Probability

If we make a bet to draw any diamond out of a well shuffled deck of cards, our target is any one of the 13 cards in the diamond suit. Therefore, we have 13 out of 52 chances. Our probability of success is $^{13}/_{52}$.

We have added together the probability of drawing any one of thirteen diamonds to get $^{13}/_{52}$. This is because any one of the thirteen cards will do.

If we make a bet to draw one card that is **either** a diamond (**13** chances out of **52**) **or** a spade (**13** chance out of **52**) from a well shuffled deck of cards, we have 26 chances out 52 to succeed. Our probability of success is: $^{1}/_{2}$. When we flip a coin, the probability that we will get **either** heads **or** tails is: $^{1}/_{2} + ^{1}/_{2} = 1$. In a coin flip we are certain to get either heads or tails.

We calculate the probability that an outcome **is any one out of a group of possible outcomes** by **adding** the probabilities of the individual outcomes in the group. The underlying assumption, always, is that the outcomes are independent of each other. In a fair lottery, one lottery ticket has the same probability of winning as another ticket for the same draw. A player who buys 10 tickets has 10 chances of winning because each ticket provides one chance. (Of course, having more chances to win does not guarantee a win. Statistics is only an analytical tool.)

23.12 Everyday use of Probability – an Example

Any lottery is a game of chance. There are two outcomes for a player: Win or No Win. If the probability of winning is: **x**, then the probability of not winning is: **1 - x**.

Let us make up a fictitious lottery game and use it as an example to analyze our chances of winning. We will call the game SureThing. It is a purely random game based on matching 5 **unique** numbers in a draw, each number between 1 and 49, inclusive. To win a player only has to match the draw. The order in which the player selects the individual numbers is not important.

Let us analyze a SureThing player's picks for a game.

The first pick comes from a pool of 49 numbers. The second comes from a pool of 48 numbers because the numbers do not repeat. The third comes from a pool of 47 numbers, the fourth from a pool of 46 numbers, and the fifth from a pool of 45 numbers.

Each of the five picks has the following probability of getting a match in the draw:

Pick number 1: $^{1}/_{49}$

Pick number 2: $^{1}/_{48}$

Pick number 3: $^{1}/_{47}$

Pick number 4: $^{1}/_{46}$

Pick number 5: $^{1}/_{45}$

Since the probability for one number to be a match is independent of the probability for another number, the probability of all five picks matching the draw is the product of the individual probabilities.

Probability of a win: $^{1}/_{49} \times ^{1}/_{48} \times ^{1}/_{47} \times ^{1}/_{46} \times ^{1}/_{45}$ = **4.370131e-9**, or roughly, **0.00000000437**. (I did the calculation on the relatively imprecise calculator on my cell phone.)

Clearly, the probability that an individual game is a winner is quite small, close to zero. If we play ten games, each game gives us one chance to win. The ten games together give us ten chances, so we add the probabilities of the ten chances. We go from **4.370131e-9** to **4.370131e-8**, still not far from zero.

The probability that a game will not be a winner is **1 - 4.370131e-9**, which is very close to **1.0**, meaning that not winning is practically assured. Remember that probability of **1.0** means an event is assured.

In other to be the only winner of a draw, all the other players must be non-winners. Assuming that each player picked independently of the others, the probability of that happening is:

probability of our win × probability of player 2 loss × probability of player 3 loss × probability of player 4 loss × ...

The chance of that happening is smaller still!

This does not mean that we should not play the lottery. After all people do win. Statistics only gives us a measure of how lucky we have to be.

23.13 Review Problems

The following are the observations of the boy's travel time that we discussed previously. We have dropped the smallest (15) and largest (30) observations and added the observation for the first school day of the fourth week to make 14 observations.

{18, 18, 19, 19, 20, 20, 20, 20, 21, 22, 23, 24, 20, 23}

i. Calculate the Arithmetic Average.

ii. Calculate The Mean Absolute Deviation

iii. A gambling club uses a loaded coin. Over the years, the probability of **heads** has been determined to be $^3/_5$ (0.60). What are the probabilities of the following sequence of outcomes?

   a. Heads-Heads
   b. Tails-Tails
   c. Tails-Tails-Tails

iv. What is the probability that John will draw a Jack from a well shuffled deck of cards, and, without putting it back, draw another Jack?

Answers to Review Problems

2.2.1 Addition - Review Problems

   i.   48
  ii.   55
 iii.   307
 iv.   1147
  v.   550
 vi.   42

2.3.1 Subtraction - Review Problems

   i.   9
  ii.   9
 iii.   11
 iv.   1000
  v.   0
 vi.   21
 vii. $55

2.4.3 Multiplication - Review Problems

   i.   56
  ii.   456
 iii.   532
 iv.   24200
  v.   240
 vi.   245
 vii.   338

2.5.2 Division - Review Problems

   i.   1
  ii.   3 (Hint: go left to right.)
 iii.   1
 iv.   2
  v.   14
 vi.   4
 vii.   3

3.6 Negative Numbers - Review Problems

   i.   7
  ii.   -2

iii.   67
iv.   -9
v.   13
vi.   Grey: 28, Red: 27

4.3 Inequalities - Review Problems

i.   False
ii.   True
iii.   True
iv.   False
v.   True
vi.   False
vii.   True

5.3 Absolute Value - Review Problems

i.   5
ii.   25
iii.   -2
iv.   -2
v.   20
vi.   -13
vii.   9

6.4 Order of Operation - Review Problems

i.   5
ii.   16
iii.   2
iv.   0
v.   24
vi.   0
vii.   10

6.5.2 Other Uses of () - Review Problems

i.   16
ii.   40
iii.   7
iv.   1
v.   16
vi.   76
vii.   -10

7.5 Introduction to Variables - Review Problems

    i.   2
   ii.   30
  iii.   4
  iv.   3
   v.   3a
  vi.

       a. 27
       b. 18

8.6 Factoring and Fractions - Review Problems

    i. $^1/_2$
   ii. $2^2/_3$
  iii. $^1/_7$
  iv. $^{17}/_{19}$
   v. $^1/_{12}$
  vi. $^1/_5$
  vii.   1, 2, 3, 6, 9, 18, 27, 54
 viii.   2, 29, 61, 71
  ix. $^1/_3$

8.7.1 Fractions - Addition and Subtraction - Review Problems

    i. $^9/_{10}$
   ii. $^{65}/_{72}$
  iii. $^{3a}/_4$
  iv. $^{3x}/_4$
   v. $2^1/_{10}$
  vi. $1^{17}/_{60}$
  vii. $1^1/_{12}$

8.8.3 Fractions - Multiplication and Division - Review Problems

    i. $^4/_7$
   ii.   1
  iii. $^7/_{12}$
  iv. $^2/_9$
   v. $^4/_{15}$
  vi. $^3/_5$
  vii. $^1/_6$

8.12 Mixed Operations with Fractions - Review Problems

    i. $2^{17}/_{60}$
   ii. $^{13}/_{15}$

iii. $1^{15}/_{16}$

iv. $^{(6ab - 3b - 2a)}/_{6ab}$

v. $^{(7b + 5ac)}/_{(7abc)}$

vi. $2^{14}/_{15}$

vii. 18 (Hint: they gave $^1/_3$ away initially, and had $^2/_3$ left. They gave away $^1/_2$ of $^2/_3$, which is another $^1/_3$. The remainder, **12**, is $^1/_3$ of their catch.)

9.9 Decimals - Review Problems

i. 316.7015
ii. 12.385
iii. 411.539
iv. 309.354
v. 0.008
vi. 1760
vii. $5.70

10.4 Rounding - Review Problems

i. 119.668
ii. 156.984
iii. 810.00
iv. 14.5612
v. 67.17
vi. 876.1
vii. 900

11.5 Ratios and Percentages - Review Problems

i. $2400
ii. 18.75%
iii. $58.54
iv. 5 feet 8 inches
v. 29%
vi. 1600

12.2.2 Simplifying Variable Expression - Review Problems

i. 6b + 4a − 73
ii. 10a + 18b + 16c
iii. 53 − 2a − 18b
iv. 10a − 18b
v. 10a + 2b
vi. 26 − 2a − 4c

12.3.1 Evaluating a Variable Expression - Review Problems

i. 1)   9

2) -33

3) 56

ii. 1) $1^1/_2$

2) $2^3/_4$

3) $^3/_4$

13.4 Single Variable Equations - Review Problems

i. $3x + 14 = 41$; $x = 9$
- **x** represents the number.

ii. $2x + x = 18$; $x = 6, 2x = 12$
- **x** is the number of fish caught by the boy with the smaller catch. the other boy caught twice as much: 12.

iii. $^1/_2x - 10 = ^1/_3(x - 10)$; $x = 40$
- **x** is the father's age today. His son is $^1/_2x$ today. The father is 40 and his son is 20 today. Ten years ago he was 30 and his son was 10.

iv. $24 \times 14 + 132 = 18x + 20(24 - x)$; $x = 6$
- **x** represents the number of shovels sold for $18. The total sale price equals the purchase price plus the profit.

14.3 Systems of Linear Equations - Review Problems

i.   Answer: **x** = 4
- Solve the equations: **y** = 7.5**x**, and **y** = 5**x** + 10, where **y** represents fee charged and **x** represents number of shrubs planted.

ii.   Answer: Company B
- The solution for the equation: **55 × 10 = 50 × 10 + .20m**, where **m** represents the number of miles the family drives, tells us the mileage at which the fees of the two companies are the same. The solution is: **m = 250**. Below **250** miles, Company B is cheaper. Since the family expects to drive **230**, they are better off with Company B.

iii.   **y** = $^1/_4$, **x** = $3^1/_4$

iv. $y = 4$, $x = 6$

15.6 Exponentials - Review Problems

 i.   181
 ii.   59
 iii. $-10^8/_9$
 iv.   4
 v.   Daughter: 5, Son: 10, Man: 50
 vi. $24

16.4 Square and Cube Roots - Review Problems

 i.   ±9
 ii.   -5, 7
 iii. 7, -5
 iv.   13
 v.   -5
 vi.   32
 vii. 18.868

17.4 Polynomials - Review Problems

Create Polynomials from their factors:

- $x^2 - x - 2$
- $6x^2 + 10x + 5$
- $6x^2 - 13x + 5$
- $6x^2 - 7x - 5$

Factors of Polynomials:

 i. $(x - 1)(x - 1)$
 ii. $x(x - 1)$
 iii. $(4x + 2)(x - 5)$

18.3 Solution of Quadratic Equation - Review Problems

 i.   The solution is: $x = 2$, $x = -2$
   - The factors are: $(x - 2)(x + 2)$

 ii.   The solution is: $x = {}^1/_2$, $x = -4$
   - The factors are: $(2x - 1)(x + 4)$

 iii. The solution is: $x = {}^1/_2$, $x = -2$

- The factors are: $(4x - 2)(x + 2)$

iv.   The solution is: $x = 7$, $x = -6$
- The factors are: $(x - 7)(x - 6)$

19.3 The Quadratic Formula - Review Problems

i. $x = 2$, $x = -2$

ii. $x = \frac{1}{2}$, $x = -4$

iii. $x = \frac{1}{2}$, $x = -2$

iv. $x = 7$, $x = -6$

20.2.1 The Straight Line - Review Problems

i. $145°$
ii. $35°$

20.3.4 The Triangle - Review Problems

i. $48°$
ii.   4.375 inches
iii.   8.60 inches

20.4.1 The Rectangle - Review Problems

i.   64 feet
ii.   56 feet
iii.   198 square feet

20.5.1 The Circle - Review Problems

i.   87.92 square inches
ii.   a) 7 inches
b) 43.96 inches

iii. $720°$

20.6.1 Surface Area and Volume - Review Problems

1)a. 16

1)b. 144 sq. inches

2)a. 9.42 cubic inches

2)b. 38.465 square inches

21.3.1 Coordinates of a Point - Review Problems

No answers provided.

21.4.1 Equation of a Straight Line - Review Problems

  No answers provided for (a - d)

   i.   C(1,4) and D(-1,-2) lies on: y = 3x + 1

   ii.   A(1,-3) and B(3,1) lies on: y = 2x – 5

   iii.   G(1,2) and H(5,10) lies on: y = 2x

   iv.   E(4,6) and F(8,8) lies on y = $^1/_2$x + 4

21.5.1 Slope of a Line - Review Problems

   i.   Slope = 2
   ii.   Slope = 3
   iii.   Slope = $^1/_2$
   iv.   Slope = 3
   v.   Slope = $^1/_2$
   vi.   Lines **ii**. & **iv**. are parallel; lines **iii**. & **v**. are parallel.

21.6.1 General Equation of a Straight Line - Review Problems

   i.   Slope: 2, y-intercept (0,4)
   ii.   Slope: 17, y-intercept (0,11)
   iii.   Slope: $^3/_4$, y-intercept (0,5)
   iv.   Slope: $^5/_2$, y-intercept (0,-12)

21.7.3 Finding the Equation of a Straight Line - Review Problems

   i.   y = 3x + 1
   ii.   y = 2x - 5
   iii.   y = 2x
   iv.   y = 1/2x + 4
   v.   y = 7x -1

21.9.1 An Application of Linear Equations - Review Problems

No answers provided.

21.10.1 Parabola - Review Problems

No answers provided.

22.2.1 The Histogram - Review Problems

    i. $90 thousand
    ii. $20 thousand
    iii. $170 thousand

23.3.1 The Pie Chart - Review Problems

    i. $505
    ii. $1313

22.4.1 The Line Chart - Review Problems

    i.   8 assists
    ii.   6 assists
    iii.   3 times
    iv.   No

23.13 Basic Statistics - Review Problems

    i.   20.5
    ii.   1.5
    iii.

        a. $9/25$ (or 0.36)
        b. $4/25$ (or 0.16)
        c. $8/125$ (or .064)

    iv.   0.0588 (to 4 decimal places)

Supplement I: Self-Assessment Test

If you studied the material in this primer in preparation for a test, like the GED math test, you are encouraged to take the following self-assessment test as part of your overall preparation. There are 34 questions. Give yourself one hour to complete the test. That gives you a little under an average of two minutes to answer each question. Some questions should take under a minute to solve, while others may take more than a couple of minutes.

The correct answers are provided at the end of the test questions. Compare them to your own answers. The result will help you to determine what topics you may still need to review further.

Test Questions

1. Mark starts with $327 in his checking account. He writes a check for $189, then a second check for $73. Which of the following operations shows how much is left in his account?

a) $189 + $73

b) $327 + $189 + $73

c) $327 − $189 − $73

d) $327 + $189 − $73

2. Four friends carpool to work Monday through Friday. Each week they pay a total of $162 for gas, parking, and tolls. Which of the following shows how they could split the cost?

a) 4 + $162

b) $162 ÷ 4

c) $162 − 4

d) $162 × 4

3. John's weekly expense for commuting to NYC is $280. It consists of gas, tolls, and parking. Parking is twice the cost tolls, and half the cost of gas. How much does John spend on tolls?

a) $50

b) 160

c) $40

d) $80

4. Given **a** = ⅓, **b** = 3, evaluate the following expression:

$$\frac{ab - a}{b - a}$$

The answer is:

a) $^1/_4$

b) $^2/_3$

c) $2^2/_3$

d) $1^7/_9$

5. Which of the following statements is True?

a) $|-2 - 3| < 5$

b) $|-2 - -7| > 5$

c) $|-2 - 3| \leq 5$

d) $|-2 - 3| = -5$

6. Evaluate the following expression and select the correct answer from the choices:

$$-2(14 - 7) + 27 \div 9 - 8 \times 7$$

a) -67

b) -43

c) -102

d) 43

7. Simplify the following expression and select the correct answer from the choices:

$$3x - 2(2x - y) + (x - y)$$

a) $7x - 3y$

b) y

c) $-3y$

d) $3y$

8. Which of the following numbers are prime?

       1, 2, 15, 17, 21, 31

a) 1, 15, 31

b) 1, 15, 17, 31

c) 1, 2, 17, 31

d) 5, 17, 31

9. A store increased the sale price of its most popular shoe by 12% before a general sale. It sold 72 pairs of the shoe during the sale for a total of $4032. What was the price of the shoe before the increase?

a) $50

b) $63.64

c) $49.28

d) $56

10. What is the **perimeter** of the blue shaded area?

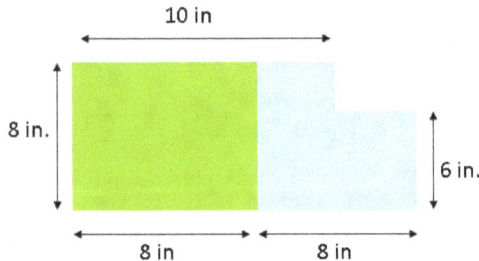

a) 32

b) 30

c) 24

d) 16

11. Referring to the following figure, what is the **area** of the blue shaded part? Round your answer to 2 decimal places.

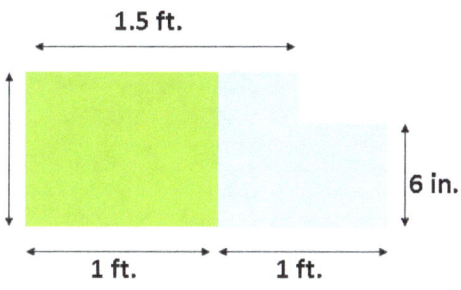

a) 7.00 sq. ft

b) 0.58 sq. ft

c) 7.00 ft.

d) 84.00 sq. ft.

12. A high school exercise program is so popular that parents attend. The current session has 100 participants, of which 30% are female. Male parents make up 10% of the rest. How many male students are in the exercise program?

a) 70

b) 60

c) 63

d) 80

13. Two men, Andy and Joe, went mountain climbing. Andy climbed 1400 ft. over 7 days. Joe climbed 735 ft. over $3^1/_2$ days and hurt his foot. What percentage of Andy's **average** daily climb was Joe's **average** daily climb?

a) 95%

b) 52.5%

c) 105%

d) 190%

14. In his math class, Josh drew 2 circles, with radius of 10 in. and 7 in. If the smaller circle is placed on top of the larger circle, what percentage of the larger circle's area will the smaller circle cover? Use π = 3.14.

a) 70%

b) 49%

c) 1.43%

d) 0.7%

15. Which of the following statements is false?

a) $(-3 + 7) = 7 - 3$

b) $\frac{1}{a} + \frac{2a}{b} = (b + 2a) \div ab$

c) $\frac{1}{a} + \frac{2a}{b} = (b + 2a^2) \div ab$

d) An equilateral Δ has $60°$ angles.

16. Reduce the fraction: $\frac{1638}{2457}$ its lowest terms. The result is:

a) $\frac{13}{21}$

b) $\frac{2}{3}$

c) $\frac{7}{13}$

d) $\frac{9}{13}$

17. What is the value of **h** in the following equation:

$$3h - 13 = \frac{h}{2} + 22$$

a) 35

b) 10

c) $3\frac{3}{5}$

d) 14

18. Today, a man is twice his son's age, and ten years more than twice his daughter's age. Five years ago the man's age equaled three times his daughters age. How old is his son today?

a) 20

b) 15

c) 30

d) 25

19. The price of gas at a local gas station is $1.98/gal. If Joshua paid $39 to fill his tank, how many gallons of gas did he buy? Round your answer to 2 decimal places.

a) 20.01 gals.

b) 19.60 gals.

c) 19.70 gals.

d) 20.10 gals.

20. The count of people at a college football game was 64,492. This number rounded to the nearest thousand is:

a) 64,000

b) 64,400

c) 65,000

d) 60,000

21. Two squirrels, Red and Brown, are playing a hopping game on a number line that is marked with consecutive integer positions. Red starts at **0**, and Brown starts at **21**. They hop simultaneously, both of them heading in the direction of decreasing numbers. Red decreases its position by **1** with each hop, while Brown decreases its position by **4**. Which of the following equations tells how many hops they have taken when they land on the same number?

a) $x = 21 + 4x$

b) $x = 21 - 4x$

c) $-x = 21 - 4x$

d) $x + 1 = 21x \div 4$

22. What is the equation of the straight that goes through the following two points?

   A(0,1), B(2,4)

a) $y = 3x + 2$

b) $2y = 3x + 2$

c) $y = 3x + 1$

d) $y = 2x + 4$

23. What is the slope of the following straight line?

$$3y = 4x + 10$$

a) $1^1/_3$

b) $^3/_4$

c) 4

d) 10

24. What is the area of the following triangle?

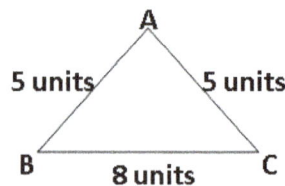

a) 12 sq. units

b) 40 sq. units

c) 25 sq. units

d) 24 sq. units

25. What are the volume and surface area of the following box?

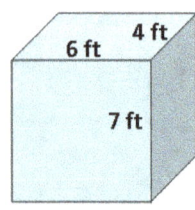

a) volume =168 cu. ft., surface area = 188 sq. ft.

b) volume = 42 cu. ft. surface area = 94 sq. ft.

c) volume = 188 cu. ft., surface area = 168 sq. ft.

d) volume = 24 cu. ft., surface area = 42 sq. ft.

26. Evaluate the following expression:

$$1^1/_5 - ^2/_5 \div ^4/_7 + 1 \times ^2/_3$$

The result is:

a) $1^1/_{15}$

b) $2^1/_{15}$

c) $1^1/_6$

b) $1^3/_5$

27. Which one of the following statements is false?

a) $\sqrt{25} \times \sqrt{25} = 25$.

b) The square of (-5) is 25.

c) $^3\sqrt{-1000} = -10$

d) $5^2 - 3^2 = 2^2$

28. The following line chart shows the number of assists the guard, Juan, on a fictitious high school basketball team had per game over a ten game period. An assist is a pass to a teammate that immediately results in a score. Between which consecutive games did Juan register his biggest increase in assists?

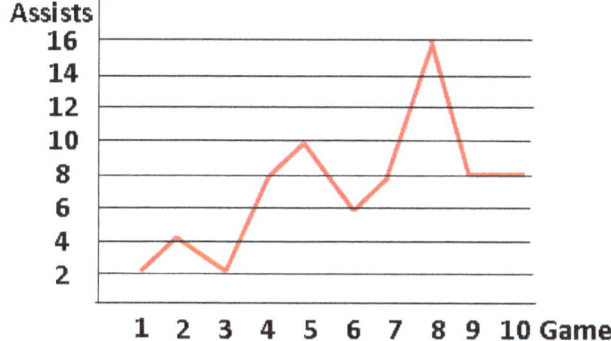

a) Game 3 and game 4

b) Game 8

c) Game 7 and game 8

d) Game 8 and game 9

29. Referring the assist chart in the previous question, what is Juan's average assist per game over the ten games? Round to the nearest whole number.

a) 7.0

b) 8.0

c) 14.0

d) 12.0

30. Again, referring Juan's assist chart, what is the **range** of his assists per game over the ten games?

a) 10

b) 14

c) 8

d) 6

31. The following pie chart shows how Mary currently spends her $2400 monthly salary. If her rent goes up 5% and she covers the increase by reducing miscellaneous expenses, how much of her salary will now go towards miscellaneous expenses each month?

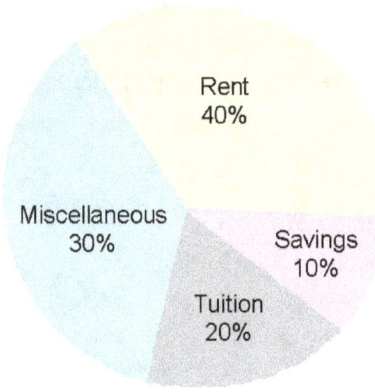

a) $720

b) $600

c) $1080

d) $120

32. The square of Emma's age is 10 years less than her father's age. If Emma's father is 35 years old, how old is Emma?

a) 6.7 years

b) 10 years

c) 3.2 years

d) 5 years

33. Which of the following statements are false?

1. $\sqrt{64}$ = -8

2. $\sqrt[3]{-64}$ = 8

3. **81** raised to the power $(-\frac{1}{2})$ equals $\frac{1}{9}$.

4. $(2\mathbf{x} - 3) - (2\mathbf{x} - 3)^0 = 2\mathbf{x} - 2$.

5. $\mathbf{a}^3 + \mathbf{a}^3 = \mathbf{a}^6$

a) Statements 3, 4.

b) Statements 1, 4.

c) Statements 2, 5

d) Statements 1, 2, 3, 4, 5

34. Mrs. Hudson has paid to have the shrubs on her property re-planted twice. The first time, Adam did it. Adam charges a flat $6.75 per shrub planted. The second time, Jerome did it. Jerome charges $15 for service call and $5.50 per shrub planted. If Mrs. Hudson paid the same amount on both occasions, how many shrubs does she have?

a) 15

b) 60

c) 11

d) 12

Self-Assessment Test Answers

1. c)   2. b)   3. c)   4. a)   5. c)   6. a)

7. b)   8. c)   9. a)   10. a)   11. b)   12. c)

13. c)   14. b)   15. b)   16. b)   17. d)   18. d)

19. c)   20. a)   21. c)   22. b)   23. a)   24. a)

25. a)   26. c)   27. d)   28. c)   29. a)   30. b)

31. b)   32. d)   33. c)   34. d)

Supplement II: Multiplication Table

×	1	2	3	4	5	6	7	8	9	10	11	12
1	1	2	3	4	5	6	7	8	9	10	11	12
2	2	4	6	8	10	12	14	16	18	20	22	24
3	3	6	9	12	15	18	21	24	27	30	33	36
4	4	8	12	16	20	24	28	32	36	40	44	48
5	5	10	15	20	25	30	35	40	45	50	55	60
6	6	12	18	24	30	36	42	48	54	60	66	72
7	7	14	21	28	35	42	49	56	63	70	77	84
8	8	16	24	32	40	48	56	64	72	80	88	96
9	9	18	27	36	45	54	63	72	81	90	99	108
10	10	20	30	40	50	60	70	80	90	100	110	120
11	11	22	33	44	55	66	77	88	99	110	121	132
12	12	24	36	48	60	72	84	96	108	120	132	144

This is how to use the table. To find the result of multiplying two numbers, follow the cells horizontally for one of the numbers and vertically for the other number. The cell at the intersection of the two paths contains the result. For example, the pink paths start on **5** and **6** and intersect at **30**, telling us that **5 × 6 = 30**. Additional multiplication results have been traced using different colors.

Acknowledgements

I thank Mr. Kwaku Danso, PhD, Mr. Scott Kuchinsky of the Plainfield (New Jersey) Library Literacy Center, and, especially, Mr. Victor Dorbu for reviewing the draft of the manuscript. Victor, I can't thank you enough.